S0-DTC-209

About Island Press

Since 1984, the nonprofit organization Island Press has been stimulating, shaping, and communicating ideas that are essential for solving environmental problems worldwide. With more than 800 titles in print and some 40 new releases each year, we are the nation's leading publisher on environmental issues. We identify innovative thinkers and emerging trends in the environmental field. We work with world-renowned experts and authors to develop cross-disciplinary solutions to environmental challenges.

Island Press designs and executes educational campaigns in conjunction with our authors to communicate their critical messages in print, in person, and online using the latest technologies, innovative programs, and the media. Our goal is to reach targeted audiences—scientists, policymakers, environmental advocates, urban planners, the media, and concerned citizens—with information that can be used to create the framework for long-term ecological health and human well-being.

Island Press gratefully acknowledges major support of our work by The Agua Fund, The Andrew W. Mellon Foundation, Betsy & Jesse Fink Foundation, The Bobolink Foundation, The Curtis and Edith Munson Foundation, Forrest C. and Frances H. Lattner Foundation, G.O. Forward Fund of the Saint Paul Foundation, Gordon and Betty Moore Foundation, The Kresge Foundation, The Margaret A. Cargill Foundation, New Mexico Water Initiative, a project of Hanuman Foundation, The Overbrook Foundation, The S.D. Bechtel, Jr. Foundation, The Summit Charitable Foundation, Inc., V. Kann Rasmussen Foundation, The Wallace Alexander Gerbode Foundation, and other generous supporters.

America's Urban Future

America's Urban Future

Lessons from North of the Border

Ray Tomalty and Alan Mallach

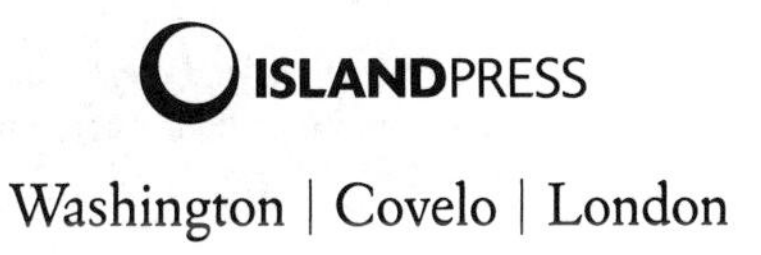

Washington | Covelo | London

Library of Congress Control Number: 2015943854

Printed on recycled, acid-free paper 

Manufactured in the United States of America

10 9 8 7 6 5 4 3 2 1

Keywords: Density, equalization, governance, immigration, land use, livability, social policy, suburban sprawl, sustainability, sustainable suburbs, transit, transportation, urban connectivity, urban development, urban form, urban policy

Contents

Acknowledgments

Many people shared their time, information, and insights with us from the very beginning to make this book possible. Its origins lie in a report that we prepared at the behest of Richard Oram, chairman of the Fund for the Environment and Urban Life (FEUL), who then encouraged us to turn it into this book. Al Cormier, a member of the FEUL board of directors, and Tom Wright and Chris Jones of the Regional Plan Association in New York provided valuable initial guidance.

In Canada, we have benefited from the insights of numerous scholars and practitioners, including Zack Taylor and Larry Bourne of the University of Toronto, Marcy Burchfield of the Neptis Institute, Caroline Andrew of the University of Ottawa, Craig Townsend of Concordia University, and Martin Wexler and Nancy Shoiry of the City of Montreal.

Kate Thompson of Dalhousie University contributed much of the material on Halifax to chapter 8, and Alon Goldstein and Keharn Yawnghwe provided invaluable research assistance. Paul Schimek of TranSystems and Michael Lewyn at the Touro Law Center helped us update some of the data they gathered for their previous work comparing US and Canadian cities. Meghan Winters of Simon Fraser University generously provided us with data on cycling facilities in several US and Canadian cities.

At Island Press, Heather Boyer was instrumental in ferrying the manuscript through its various stages and helping us organize the figures and tables, much of it redrawn by Marce Rackstraw.

Finally, as with any such project, this book would not have been possible without the constant support and encouragement of our partners, Michele Luchs (Ray Tomalty) and Robin Gould (Alan Mallach).

Introduction

Today is a time of crisis and opportunity for urbanism in the United States. By 2050, the United States will have to find room to accommodate 75 million more people, spend trillions to repair and replace deteriorating and often obsolete infrastructure, and adapt cities and suburbs alike to the near certainty of massive climate change. How we address these challenges will not only determine the course of growth for both central cities and suburbs in the United States, but will largely determine the physical environment, quality of life, and standard of living for what will be a nation of 400 million people. As we move forward and try to figure out how best to address these challenges, we need—to put it bluntly—all the help we can get. Although there are many good models of urban planning, smart growth, and urban transportation within the United States, there are many more in other countries. We need to learn whatever we can from them about what they offer to help guide thinking and action in this country.

We are not the first to come up with this idea, and we will not be the last. Over the past decades, many scholars and practitioners have looked at other countries with generally similar levels of development, that is, highly urbanized and postindustrial countries like Canada, Australia, and in particular, the advanced economies of Western Europe. We believe that our book, while standing squarely in that tradition, breaks important new ground. We believe that a comparison of the United States and Canada should be particularly useful, arguably more so than more widely discussed and disseminated European models, whose relevance for the United States tends often to be rather more aspirational than practical.

There is no question that European examples of urban sustainability and the robust policy initiatives that have given rise to them should be of interest to American audiences, but we share a lingering doubt about their relevance.

Cities of the Old World are simply too different from those of the United States, as are European governmental institutions, legal principles, and systems of taxation. Canada offers what may be a more useful model, one close enough to be comparable yet different enough to be instructive.

Why Canada?

Cities in Canada seem to fall somewhere between cities in Europe and the United States on most indicators of urban livability and sustainability. They are extraordinarily similar to cities in the United States in their spatial system and built environment. They typically have an older core, usually based on a grid street system, with the tallest buildings clustered in a historically largely nonresidential downtown, surrounded by inner-core residential areas. This core is surrounded in turn by industrial areas, many of which are today either fallow or redeveloped for other uses, and relatively tightly knit older suburbs. Beyond them in turn are the newer suburbs, forming a steadily expanding ring of low-density, car-dependent, sprawling development. Highways crisscross the region, linking central cities with suburbs, ringing the suburbs, and linking it with other regions, producing an interconnected system of limited-access, high-speed roads. Public transit infrastructure, particularly fixed-rail transit, in most North American cities is modest compared with that of many European cities.

Canadian cities and regions have been subject to many of the same larger forces—including rapid population growth, large-scale immigration, historically inexpensive energy, abundant land resources, and a strong market economy—that have shaped US cities. The United States and Canada are akin as well in their governance structures, including the basic federal system of government, and a common planning system grounded in comprehensive or master plans, subdivision control, and zoning. All these factors suggest that the Canadian experience is likely to be highly relevant to conditions in the United States.

A closer look reveals important distinctions, however. Most Canadian central cities have core areas that are lively around the clock. In those areas, people generally drive less, use transit more, and are more likely to be seen riding bicycles on city streets than in similar US cities. Most older suburbs are still highly desirable communities, and newer suburbs are more compact and better served by transit. Urban-suburban social and economic disparities are less pronounced and, in many cases, nonexistent. Both racial segregation and economic segregation are less prevalent and rarer in Canadian cities than in US cities, families with children are as likely to live in central cities as in suburbs, and urban schools are considered as good as their suburban counterparts.

Canadian cities and metropolitan areas, or metros, are no more homogenous than are those in the United States. Some cities like Toronto are compact with many walkable neighborhoods and extensive transit systems, whereas others like Calgary and Saskatoon are less compact and relatively car-dependent, just as Portland, Oregon, can be similarly contrasted with cities like Dallas or Houston. We are *not* arguing in this book that Canadian cities are paragons of smart growth virtue. Canadian cities suffer from many of the same ills as their United States counterparts. Growth in fringe areas outweighs growth in the already urbanized areas, much commercial activity takes place along car-dependent arterial strips, power centers and big-box stores have multiplied, automobile ownership is high, and congestion is a serious problem in larger urban centers. Meanwhile, much employment has shifted to suburban office parks often poorly served by transit.

Our point is not that *every* Canadian city is different from *every* city in the United States; despite the seemingly powerful evidence for similarity, however, we find that on the whole, Canadian regions are far more likely to show a more compact, less car-dependent, profile and that their central cities are more likely to be vital, thriving entities than are their US counterparts. What is the norm in Canada is the exception in the United States. Portland, Oregon, is an outlier in the United States, but would be far closer to the norm in Canada.

Given the two nations' many institutional, cultural, and economic similarities, one must wonder why their cities and regions have diverged to such an extent in terms of urban form, reduced car dependency, and livability. Although there are cultural differences between Canadians and Americans, as we discuss in chapter 2, we believe that those are less important than some observers have suggested, except—and we do not underestimate this point—as they serve to provide a value-based underpinning for policy differences. It is those policy differences, however, that interest us: how each nation—its states or provinces and its individual municipalities—has attempted to shape growth and development and how those differences lead to different outcomes in terms of livability and sustainability. We explore those differences in detail and then trace how they have led to differences in the urban form and the course of development in the two countries. Finally, we draw some lessons from our exploration to inform potential efforts for policy reform in the United States.

Our Starting Point

We are concerned about the future of our cities and regions, both in the United States and Canada, and about their livability and sustainability as they confront the challenges of the coming years. Those terms are fraught with political and social as well as economic and environmental implications, and we should make our own perspective on those terms—and their relationship to urban growth and change—clear.

We are not antisuburb or anticar. Suburban development, beginning in the late nineteenth century, has enabled millions of families in the United States and Canada to live better lives, and the automobile has given millions of people an unprecedented level of mobility and opportunity. At the same time, we believe that both suburbanization and automobile dependency carry with them dangerous baggage. The line between suburbanization and sprawl is a fine one, and it is not always easy to tell until long after it has been crossed. As regions sprawl, some may gain; much, though, is lost, not only in terms of excessive, wasteful consumption of land, energy, and resources, but in terms of the disconnection of the parts from the whole and an increasing inequality of resources and opportunity within the region. Automobiles are invaluable for some purposes, but car dependency, particularly for the journey to work, has triggered unsustainable levels of energy consumption, congestion and disruption of quality of life. These concerns go directly to both livability and sustainability.

For all the much-vaunted quality of life in many United States metropolitan areas, we, along with many of our colleagues, find ourselves deeply concerned about many of its manifestations. As we look at our regions as they have spread across the landscape, sprawl has fueled increasing spatial barriers to opportunity and increasing disparities of social and economic condition. Ghettoization of the poor and minorities is unabated and, indeed, has spread to many of the modest postwar inner-ring suburbs, the first destination of white flight in the 1950s and 1960s. Despite signs of revival in downtowns and around universities, older cities in the United States remain deeply distressed, and the phenomenon of "sprawl without growth" as land consumption at the fringe exceeds regional population and household growth continues. Our ideal of a livable region is one in which all residents can enjoy a decent quality of life and can find opportunity, without regard to their income, family status, or where they fit into the spatial system of the region. Although no region fully meets that test, those in the United States fall far short, appallingly so, in light of our resources as a nation.

We share the same concerns about the sustainability of the growth pattern that the United States has followed over recent decades, in particular the profligate consumption of land, resources, and energy associated with growth and the potentially dire consequences of that for our future and that of our children and grandchildren. By definition, a sustainable course of growth is one that will enable not just North Americans, but people throughout the world, to live a decent quality of life that can be sustained over the long term within the limits of the planet's resources and its natural environment. One cannot look at our world today without deep concern that we may have already failed; the evidence is increasingly compelling that we are creating long-term and potentially disastrous changes in our natural environment, while billions still live in poverty and want.

This book will not affect the course of those global issues. We hope, though, that it may contribute modestly but meaningfully to incremental policy change in the United States with respect to smarter, more sustainable future growth, development, and redevelopment of its central cities, suburbs, and metropolitan areas.

On the Shoulders of Giants

Although our book contains much that is new and different, we are far from the first writers to compare the United States and Canada. Interest in comparison of the two countries has been a recurrent theme in academic and policy circles for a long time, although clearly more so in Canada, where the presence of the United States is overwhelming and inescapable, than in the United States. That interest has spawned a series of books comparing the history, political structure, economic systems, cultural formations, and value systems in the two countries, including a widely used anthology, *Canada and the United States: Differences That Count,*[1] now in its fourth edition since it initially appeared in 1993; and two widely read more value-oriented studies, consultant Michael Adams's *Fire and Ice: The United States, Canada and the Myth of Converging Values*[2] and prominent sociologist Seymour Martin Lipset's *Continental Divide: The Values and Institutions of the United States and Canada,*[3] perhaps the only book in this genre written by an American author resident in the United States and published by a US firm.[4]

Scholarly interest specifically in the similarities and differences between cities in the United States and Canada goes back to the publication of *The North American City* by Maurice Yeates and Barry Garner in 1976.[5] The book, intended as an undergraduate geography textbook, looked at the system of cities in North America, analyzed the structure of urban areas, and applied the authors' findings to public policy questions. Replete with charts and graphs and written in an accessible style, *The North American City* received wide circulation.

The book was widely challenged for its failure to differentiate Canadian from US cities, treating them largely as the products of common historical and geographic trends. The most influential response to Yeates and Garner came from economist Michael Goldberg and geographer John Mercer in their 1986 book, *The Myth of the North American City: Continentalism Challenged.*[6] Based on rigorous quantitative analysis of a large collection of relevant variables, the authors concluded that "continentalism," as they call it, was deeply flawed:

> *Overall, the ... analyses generally support the contention that Canadian cities are sufficiently different and distinctive within a North American context that they require separate consideration. While Canadian and United States cities may be subject to similar causative processes, such as the transformation of employment structures, population deconcentration or immigration, there are other processes which are structured*

> *differently and perform differently, such as intergovernmental relations.... Canadian urban areas are very different places to those in the United States. Hence, the notion of the "North American City" can be of only limited value and may be potentially misleading.*[7]

The Mercer and Goldberg thesis, including the authors' heavy reliance on cultural explanations for the differences they observed, was not universally accepted. Frances Frisken argued that the source of the differences was to be found in institutional and political systems and that those differences were eroding by the 1980s.[8] Other writers also saw the historic differences, whatever they might have been, being eroded by the forces of globalization.

Mercer has continued to work on this issue. Updating his analysis, he has insisted that commonalities should not be mistaken for the erosion of cross-national differences and that pressures for convergence from globalization or otherwise do not necessarily lead to common policy responses, a conclusion that we share.[9] The discussion, however, continues, reflected in a special issue of the *International Journal of Canadian Studies* in 2014 on the theme of "Reopening the 'Myth of the North American City' Debate" and a major contribution to the topic in 2015 in a brilliant dissertation from the University of Toronto's Zachary Taylor.[10]

In parallel with the "North American City" debate, a considerable number of shorter studies have appeared that compare United States and Canadian cities and metros with respect to land use planning or the extent of sprawl;[11] and with respect to transportation systems, including bicycling[12] and use of public transportation.[13] In short, in looking at the question that we have posed for ourselves in this book, there is no shortage of material to draw upon.

We have taken this material, as well as our own analyses, and done something fundamentally different with it. We examine how and why Canadian cities and metros differ so significantly from cities in the United States with respect to livability and sustainability as well as what the implications of those differences might be for future policy choices in the United States. In so doing, we concentrate on two issues that we see as particularly salient: urban form, including land use planning, development, and infrastructure; and transportation. Although we inevitably touch on many other themes, such as the social policy factors that disproportionately affect the vitality of central cities, those are the central themes of the book and they in turn form the basis for the policy recommendations in the final chapter.

We recognize that it may be a difficult time to present a body of recommendations, which in many respects can be summed up as "plan more, and better," in the United States. The policy climate, at least at the national level, is bleak, with a Republican Congress and a Democratic president deadlocked in many major policy areas, not least of which are renewable energy and climate change. At the same time, that is not the entire story. Cities are being transformed by migration of young highly educated adults, and historically car-dependent cities like Houston, Texas, and Tucson, Arizona, have built light rail lines as they try to turn their downtowns into higher-density, mixed-use places. The times are changing, and we believe that there is a thirst in the United States—perhaps not in every part of the country, but in many places—for policies that can begin to turn the many one-off, scattered, transformative efforts into systemic change. It is with this hope that we write this book.

The Plan of This Book

In the following ten chapters, we provide a systematic look at the differences between the United States and Canada, the most important forces and features that have led to those differences, and their implications for both central-city and suburban livability and sustainability. Early chapters set the stage. In chapter 1, we ask the question, why do these issues matter? We explore the major trends taking place in both the United States and Canada with respect to demographic and economic change, travel and settlement behavior, and attitudes and values, in two respects: forces that are *driving* change and factors that are fueling the *demand* for change.

Chapter 2 is something of a Canada primer, written for readers in the United States who may be relatively unfamiliar with their northern neighbor. Here we highlight both the similarities and the differences between the two nations with respect to their history, legal and political systems, regional differences, and economic and social conditions, thus providing a frame for the rest of the book.

In chapters 3 and 4, we join the "North American city" debate, the long-standing argument over whether cities and metropolitan regions in the two countries are fundamentally similar or fundamentally different. Although we recognize that cities and metros in the United States and Canada all fall along a single continuum with respect to their livability and sustainability, these chapters identify significant differences between cities in the two countries. We suggest that Canadian cities on the whole tend to fall more on the more sustainable side of the continuum and cities in the United States on the other. We also suggest that, in important respects, Canadian cities better fit the emerging vision of compact, transit-oriented, and socially inclusive places than their US counterparts.

Chapters 5 through 7 explore some of the most important factors that may account for the differences in urban function and form. Although some authors have focused on value differences to explain these differences, we focus on policy differences. That is, we focus on the levers of governmental action at all levels—federal, state/provincial, regional, and local—that lead to different outcomes and that in turn can prompt thinking about policies that might deflect the path of urban revitalization and suburban growth in the United States into more sustainable directions. In these three chapters, we look at differences in the distribution and organization of governmental powers, land use policies, transportation policies, coordination of transportation and land use planning, fiscal issues, and social inclusion policies.

In chapters 8 and 9, we pull together the various policy threads we unraveled earlier. We look at how they influence the health and vitality of central cities and affect the sustainability and livability of suburban growth, and we use case studies of selected cities and suburbs in the United States and Canada to demonstrate how these policies actually play out.

In chapter 10, the concluding chapter, we ask what these trends mean for policies and practices in the United States. We begin by offering our assessment of the policy climate for change in the United States and suggest that, despite many difficulties and obstacles, there are real opportunities for moving policy forward to reflect many of the changes taking place on the ground. From that starting point, we explore how policy changes can have an effect in three distinct areas: suburban greenfield development, suburban infill and intensification, and urban revitalization. We also offer specific proposals in a range of areas that can potentially change policy and practice in ways that will further greater livability and sustainability.

To repeat the words with which we opened this introduction, by 2050, the United States will have to find room to accommodate 75 million more people, spend trillions to repair and replace deteriorating and often obsolete infrastructure, and adapt cities and suburbs alike to the near certainty of massive climate change. Any one of those realities should be enough to make us think seriously about changing our modus operandi. All three, taken together, are a rousing call to action.

Chapter 1
Changing World, Changing Cities

Cities in the United States are in a time of transition. Post–World War II patterns of growth have begun to play themselves out, and new patterns are emerging. Although low-density, car-dependent development on the urban fringe continues, more emphasis is being placed on higher-density, mixed-use development around transit stations, in city centers, and in suburban subcenters. In a reversal of historical trends that saw middle-class people flee distressed urban cores, many central cities are attracting new residents (especially younger people). In both central cities and suburbs, Americans are increasingly demanding more walkable and transit-friendly neighborhoods and in general are looking for higher quality urban places in which to live, work, and play. Concerns over the environmental and public health effects associated with urban sprawl, tighter government infrastructure budgets, the emergence of the creative economy, and a growing awareness of the destructive implications of social inequality are causing community leaders to question conventional models of urban growth and development. All these issues will almost certainly continue to gain in importance in the coming years and contribute to this new phase in the evolution of US cities.[1]

During this time of change, urban leaders in the United States are looking for direction. Across the nation, city officials, planners, developers, architects, and others involved in shaping our cities are experimenting with new approaches to city design. Plans and projects going under a variety of rubrics—from pedestrian pockets, transit-oriented development, and complete communities to life-cycle neighborhoods and new urbanist developments—are springing up across the country. The truth is, however, that even though a number of organizations are working to spread these emerging practices by disseminating knowledge and experience to city builders around the country, these promising trends remain sporadic and scattered. Pockets of change are visible here and there, but the larger governance and policy arrangements that favor sprawl continue to churn out low-density, car-dependent development in urban regions across the country.

Within this context of change and brakes on change, our purpose is to bring a fresh perspective on US planning and development trends by leavening the discussion with experiences and practices from the country's neighbor to the north, Canada. Anyone who has visited Canada after living in the United States has experienced the uncanny "same but different" feeling. Canadian cities look more or less like US cities, with similar downtown skylines, road patterns, and architectural forms, but some differences are immediately apparent: city centers are generally livelier, even people who can afford cars take transit, there are more people on bikes, districts of concentrated crime and extreme poverty are relatively rare, there are few gated communities, and most people feel (and are) safe to walk the city streets, even at night.

Beyond these tangible differences in the urban quality of life between the two countries are less visible but no less important differences in environmental sustainability. By international standards, both US and Canadian cities do well in terms environmental conditions that affect public health in that the quality of and accessibility to potable water is high, urban air pollution is relatively manageable, access to public parks is good, and contaminated soils tend to be handled properly. On measures related to resource consumption and waste generation, however, Northern American[2] cities are among the worst offenders in the world. Even compared with other rich countries, they have very high levels of greenhouse gas emissions, energy use, solid waste creation, and water use. Although it is understandable that many observers would lump Canada and the United States together as resource gluttons, there are important differences between the two countries. For example, on a per person basis, Canadian cities tend to use substantially less energy, emit fewer greenhouses gases, use less water, produce less garbage, and release fewer contaminants into the air than do their US counterparts.

Although it is true that many factors can be adduced to explain these differences in the way cities work in the two countries, there is little doubt that the built form of the city plays an important role. Built form refers to the city's physical shape and texture, size, and underlying infrastructure. Much evidence has been amassed in recent years to show that urban form is a key determinant of urban function and in particular that the density, urban structure, and mix of land uses can exert a powerful influence on how the city works in terms of its environmental and social functionality.

Sure enough, if we take the trouble to look under the similar skylines, we see significant differences in the built forms of Canadian and US cities. In Canadian cities, urban and suburban densities are typically higher, mixed-use areas are more widespread, scattered exurban development is less common, there are fewer expressways running directly through the cities, development is more focused around transit, and urban transit services themselves tend to be of higher quality than in US cities.

This book explores these national differences in built form, how they arose, and how the Canadian experience could contribute to the discussion on building more sustainable and livable cities in the United States. The book is based on the assumption that—given the similarities in cultural, economic, and political conditions—what worked well in Canada is at least worth considering in the United States. Through the resulting policy recommendations, we seek to stimulate multilevel discussion in the United States on how to remove barriers to and otherwise encourage more compact, mixed-use development, redevelopment within existing built areas, and a slowing of sprawl on the growing fringe of our cities. Our hope is that the discussion will help trigger policy changes that take advantage of the new drivers of urban change and hasten the transition to the next American city.

Drivers of Change in US Cities

To position the discussion of similarities and differences between Canadian and US cities, we highlight some of the forces that are already creating the conditions for significant changes in how US cities are designed and built. The urban landscape in the United States is undergoing profound change as it is buffeted by new forces to which it must adapt. These drivers include emerging demographic and economic changes that are shifting housing markets, sharpening environmental and health concerns, changing expectations for the quality of the places we inhabit, and tightening constraints on government spending. These factors are bringing the issue of development and transportation patterns in the United States to a head and causing us to rethink how cities are planned and built. Generally speaking, these forces are pushing US cities in the direction of more sustainable and livable environments, and their efforts could be amplified by selective policies that have shown to be an effective framework for urbanization in our neighbor to the north.

Demographic Changes

One factor driving change in the type of housing and neighborhoods people want to live in is related to demographics. Demographer Arthur Nelson has projected trends in aging to 2030 and linked them to expected changes in household formation, household size, and household structure as well as location and neighborhood preferences.[3] His projections are based on the advent of four roughly equal-sized generational subgroups—baby boomers, Generation X, Generation Y, and millennials—with two generations, baby boomers and Generation Y, being especially important in the evolving housing market over the coming decades.

The baby boom generation, born between 1946 and 1964, now accounts for 82 million Americans and will comprise 64 million people in 2030. With the leading edge of the boomers now reaching retirement age, most observers expect to see major changes in their choice of housing type and location in that time span. Already, few boomers have children still living at home, and as more boomers lose their ability to drive and walk long distances, many are expected to trade in oversized, socially isolating, and car-dependent suburban homes. If such options are available in their communities, they are likely to seek out single-family homes on smaller lots, townhouses, and condos in or near burgeoning suburban town centers; others may be drawn to more urban areas with convenient transit linkages and good public services like libraries, cultural activities, and health care.

Because boomers are such a large part of the population and will be selling their homes, often to relocate to rented premises, there may be a glut of ownership housing and stiff competition for rental housing. Nelson estimates that about half the boomers will want to live in walkable, transit-friendly neighborhoods.[4] Although some analysts believe this estimate to be overly optimistic, there is a strong possibility that aging boomers may help shift the US housing market toward smarter forms of development.

Generation Yers, born between 1981 and 1995, now number 65 million and will grow (via immigration) to 71 million by 2030. This group will also help revolutionize the housing market in the United States. Many of these people, who will be between thirty-five and forty-nine years old in 2030, may not wish to emulate their parents' suburban lifestyle. According to Nathan Norris, many Gen Yers prefer densely settled areas where they can take full advantage of social networks, have easy walking access to daily destinations, avoid car ownership, and find rental accommodations to suit their footloose lifestyles.[5] Of course, some Gen Yers will occupy suburban housing being vacated by their parents' generation, but indications are that many will be drawn to transit- and walk-friendly communities. Rather than flee to the suburbs, some analysts believe that this generation is more likely to plant roots in walkable urban areas, exerting pressure on local governments to improve urban school districts.

The outcome of these large demographic changes could entail a greatly increased demand for something that the US housing market is not currently providing: small one- to three-bedroom homes in walkable, transit-oriented, economically dynamic, and job-rich neighborhoods. Based on his demographic research, Nelson estimates that by 2030, one-fourth to one-third of US households will be demanding the type of housing options that retail corridors and subcenters can provide. Such demand will mean increasing the supply of housing that meets this description from its current stock of 10 million to 25 million by 2030. "In effect," he says, "if all new homes built in America between 2010 and 2030 were built in those locations, demand for this option would still not be met."[6] Meanwhile, Nelson expects a large oversupply of large-lot housing to accumulate.[7]

Lower Real Incomes and Prospects for Home Ownership

Many scholars draw a direct link between rising US income following World War II and the ascendance of sprawl.[8] In a nutshell, higher incomes mean that more households can afford more land, a larger home, and more cars, making car-dependent, low-density suburbs a reflection of general prosperity. Statistical modeling has confirmed this effect for the United States. Robert Margo found that 43 percent of suburbanization that took place over the 1950 to 1980 period in the United States could be attributed to people getting richer.[9]

In the heyday of urban sprawl, from 1950 to 1970, mean family incomes rose rapidly, about 3 percent annually. Although incomes continued to rise in remaining decades of the twentieth century, the rate of increase was lower, from 1 to 2 percent annually. The first decade of the twenty-first century was the first in the post–World War II period to witness an actual decline in real incomes, with all quintiles worse off in 2010 than in 2000.[10] Moreover, by 2010, the median net worth of US households had dropped to its lowest level since 1969. Although asset prices rebounded from 2010 to 2013, median wealth did not, and the median net worth of US households in 2013 was even lower than in 2010.[11] Median secured debt, which includes real estate and auto loans, rose to $91,000 in 2011, a 30.5 percent increase from the median $69,749 owed in 2000.[12]

The weakened financial condition of US families will undoubtedly undermine their ability to afford housing in the coming years, a factor that should place more emphasis on smaller homes, smaller lots, older neighborhoods, and rental housing, all consistent with more urban living. Meanwhile, the prospect of worsening congestion, longer commuting lengths, and rising fuel prices may dissuade many people from seeking cheaper housing in exurban areas far from work opportunities. Institutional factors also point in this direction. For example lending institutions have raised the bar on mortgage qualifications by requiring higher credit scores, work histories, and down payments, which will also tend to favor smaller homes and rental units.

Concern for Climate Change

Climate change has moved to the mainstream of public consciousness and caused many people to accept the necessity of a more concerted approach to managing urban growth. We are already seeing the effect of climate change (e.g., unexpected shifts in weather patterns; the number and severity of storms, floods, droughts, and major forest fires), and further changes are expected to stress physical, social, and health infrastructure. Given the increasing media attention afforded this issue, it is not surprising that the vast majority of Americans[13]—even a majority of Republicans[14]—now believe that the climate is changing and that steps need to be taken to mitigate these changes. Other indications that climate change is looming as a key public policy issue for the coming years include that the financial sector[15] and defense establishment[16] increasingly see climate change as a real threat to economic health and national security.

Next to China, the United States is the largest greenhouse gas–emitting country on the planet (and has four times the per capita emissions rate than China). That is not due only to its status as a rich, consumerist, midlatitude country; it also reflects the nature of its cities. Other countries with similar levels of economic wealth have much lower per capita emissions.[17] One reason European countries have lower emissions is that their cities are much more energy efficient than US cities.[18]

Although there is still some debate in academic circles on the effect of sprawl on greenhouse gas emissions, there is little doubt that low-density, auto-dependent sprawl consumes more energy for heating and cooling buildings and for urban transportation than more compact development.[19] For example, one report prepared for the US Environmental Protection Agency showed that a home's location relative to transportation choices, its size, and its energy efficiency can have a huge effect on energy consumption; for example, an energy-efficient multifamily unit in a transit-friendly location consumes about one-fourth the total annual energy of a standard single-family, detached household in a conventional subdivision far from transit services.[20] An Urban Land Institute publication, *Moving Cooler*, outlined various transport-related greenhouse gas reduction scenarios based on how aggressively we change land use and associated travel patterns. The authors estimated that modeling future growth on current best practices could achieve a 20 percent reduction in emissions. A maximum effort, including comprehensive growth boundaries, minimum required densities, and jobs and housing balance as well as non–land use strategies could reduce emissions by 60 percent.[21]

With urban transportation and household heating and cooling accounting for about half of greenhouse gas emissions in the United States, it is difficult to envision a "solution" to the climate crisis that does not involve more compact, mixed-use neighborhoods that are more walkable, bikeable, and transit-friendly.

Importance of Place and Quality of Life

Since the turn of the millennium, there has been a major shift in our appreciation for the importance of quality places in the economic success of a region. Cities are gradually learning that a high quality of life and walkable neighborhoods can attract and retain workers in the knowledge-intensive and creative fields.[22] The presence of mobile "creative class" workers in turn attracts firms looking for high-quality talent, which in turn boosts the rate of cultural and technical innovation. This cycle represents a new way of thinking about urban competitiveness, which traditionally focused on the ability of cities to attract firms by offering cheap labor, good infrastructure, and a compliant local government ("positive business climate").

In his book *The Rise of the Creative Class*, Richard Florida contends that professionals, artists, and high-tech workers choose to live in cities that offer a variety of good job opportunities, an open and tolerant culture, and—more importantly to our argument here—high-quality urban amenities such as appealing natural environments and vibrant neighborhoods. He also argues for making smarter use of both our urban and suburban spaces, including higher residential densities, more mixed-use development, the infilling of suburban cores near rail links, and new investment in rail.[23] Although Florida's methods and conclusions have been criticized by other urban geographers,[24] his ideas have nonetheless permeated into the thinking of government and business leaders. Cities throughout Northern America have been scrambling recently to find the magic formula to attract the footloose but vaunted creative class.

Infrastructure Deficit and Fiscal Austerity

Since the turn of the twenty-first century, the combination of soaring military costs in the wake of the September 11, 2001, terrorist attacks, economic crisis, and slow recovery has contributed to serious budget deficits at all levels of government. Although the US economy is recovering, it is expected to remain soft, and tax receipts may remain depressed for several years or longer. As a result, federal, state, and local governments are seeking ways to curb wasteful spending and improve the efficiency of public services. Public officials are being forced to consider not just short-term budget cuts but also policy reforms that will lead to long-term efficiencies.

In this environment, it is inevitable that how communities grow and how they invest public dollars will get another look. One obvious problem is infrastructure spending. At present, about $375 billion per year is being spent by governments in the United States on new urban infrastructure and repairing existing stock. Despite this massive investment, cities cannot keep up with the need to replace existing infrastructure as it ages or meet the demand for new infrastructure as they grow. Periodic surveys of municipal infrastructure needs have shown how the infrastructure deficit has mushroomed. According to the American Society of Civil Engineers, the cost of repairing the country's roads, bridges, water systems, transit systems, and other infrastructure grew from $1.3 trillion in 2001 to $3.6 trillion in 2013.[25] This growing deficit is reflected in deteriorating infrastructure in older areas and underserviced new subdivisions. Consequences can vary from collapsed bridges, service interruptions, underground leakage of potable water, substandard sewage treatment, congested roads, inadequate transit systems to other failures with important environmental, health, social, and economic dimensions.

The reasons for the US infrastructure deficit are complicated, but suburban sprawl is increasingly seen as an important contributor. By modeling different growth patterns at the regional and state levels, Robert Burchell, one of the best-known researchers in this field, has assessed how the costs of infrastructure are linked to sprawl. These studies suggest that solid savings, usually on the order of about 20 percent, could be achieved by building communities in a more compact, contiguous pattern.[26] At the national level, Burchell and his associates calculated that compact (compared to business as usual) growth patterns could reduce twenty-five-year road-building outlays 12 to 26 percent and could see water and sewer savings of about 6.6 percent. The national tabulation put the infrastructure differential between sprawl and smart growth at more than $100 billion over twenty-five years, for a savings of about 11 percent.

These savings were achievable with only modest changes to urban form—a 20 percent increase in residential densities and a 10 percent density increase for nonresidential uses. Greater savings are achievable with more substantial shifts in urban form.[27]

These findings make intuitive sense because much municipal infrastructure is linear in nature. The more densely developed and contiguous communities are, the fewer feet of highways, local roads, sidewalks, water pipes, sewers, and utility cables they need to service a given number of residents or employees. Even so-called soft infrastructure, such as police stations and recreational centers, can be sensitive to differences in urban form because they have limited catchment areas and because new facilities must be built to serve sprawling neighborhoods effectively. There are reams of literature on this topic, and different studies report different levels of potential savings by moving from a sprawl to a smart growth model. There is little doubt, however, that compact development models result in lower infrastructure costs.

One positive corollary of the infrastructure squeeze is that community leaders are increasingly looking for ways to manage growth in more cost-effective ways. In place of the traditional reflex to expand communities to accommodate new residents and employers, there is a growing interest in managing new development to fully exploit existing infrastructure. This is a major factor behind the growing interest in redeveloping abandoned or underused areas such as former industrial properties or declining shopping centers and prioritizing other forms of infill growth. Avoiding leapfrog development and boosting the densities of new suburban development are also being seen as ways to address the infrastructure squeeze and achieve long-term savings through better management of growth.

Public Health Concerns

In recent years, increasing attention has been paid to the public health implications of urban sprawl and the potential solution offered by smarter patterns of growth. Low-density, car-oriented suburban landscapes discourage commuting by bike or foot, limit transit use (which usually involves walking to and from transit stops), and force residents to spend more time sitting in their cars.[28] As three public health physicians, Andrew Dannenberg, Howard Frumkin, and Richard Jackson, wrote in their book, *Making Healthy Places*: "The trouble is that in the last half century, we have effectively engineered physical activity out of our daily lives. Health is determined by planning, architecture, transportation, housing, energy, and other disciplines at least as much as it is by medical care.... The modern America of obesity, inactivity, depression, and loss of community has not 'happened' to us; rather we legislated, subsidized, and planned it."[29]

As it contributes to car use as a transport mode and lengthens trips, sprawl is linked to air quality problems and associated health effects: respiratory diseases, heart attack, and premature death, especially in vulnerable young children and older adult populations. Measures that reduce per capita vehicle travel are likely to limit human health risks. Increased population density, land use mixing, and street connectivity have been shown to reduce air pollution from traffic.[30] Short motor vehicle trips in urban conditions have relatively high per mile pollution emission rates due to cold engine starts and congestion, so reductions in such trips provide relatively large emission reductions. These short trips are also the trips most likely to be replaced by walking and cycling if land use patterns become more walkable.[31]

Sprawl is also linked to other public health threats, including water quality problems,[32] traffic fatalities,[33] psychological stress from traffic noise,[34] loneliness, and depression.[35] What brought urban development patterns to the fore in recent years, however, is the mounting evidence that sprawl contributes to the obesity epidemic that is now ravaging the United States. To date, a number of scientific studies have investigated the relationships between urban form, transportation, and obesity, physical activity, and associated diseases (hypertension, heart attacks, diabetes, etc.). This research has consistently found that sprawled land use correlates with increased time spent in cars and with a higher likelihood of sedentary, overweight, and obese residents.[36] For example, looking at one hundred metro areas across the United States, Roland Sturm and Deborah Cohen correlated a sprawl index with sixteen different chronic diseases, including overweight-related conditions (e.g., hypertension), respiratory ailments (e.g. emphysema and asthma), and other disorders such as abdominal problems and severe headaches. The sprawl index was found to be a significant predictor of the number of chronic medical conditions in a population.[37]

Research suggests that residents of smart growth communities achieve more of their recommended minimum requirement for physical activity through daily walking and cycling. In 2001, Keith Lawton compared average daily minutes of travel by automobile, transit, and walking by residents of Portland, Oregon, neighborhoods.[38] Although the average time spent traveling is similar for the three neighborhood types, residents of the most urban neighborhoods walked an average of 11.8 minutes daily, much more than the 3.3 minutes walked daily by residents of the least urban neighborhoods. Programs to promote physical activity through gym memberships and school activities and other interventions have met with only limited success. Many experts believe instead that building the opportunity to be physically active into daily routines, through active transport and access to recreational opportunities, is the most effective way to improve community fitness.[39] Smart growth policies are increasingly cited as a key way to enhance fitness and health, both by increasing daily walking and cycling and reducing time spent being sedentary in cars.

The Need to Address Social Equity Issues

The Great Recession of 2007–2009 and its aftermath brought into relief the immense and growing differential in wealth and life opportunities that are dividing communities in the United States and the importance of social equity in city planning. This growing awareness has been heightened by recent research showing that not only do more equitable countries have longer economic growth spells, but that the same dynamic applies at the city level as well. It seems that income inequality, racial segregation, and political fragmentation within US metropolitan regions can interrupt growth, an alarming fact in an era when cities are struggling to compete in an increasingly globalized world.[40]

Of course, a certain degree of income inequality is expected in any metropolitan region. To some extent, people tend to filter themselves into neighborhoods and municipalities based on housing costs, transportation opportunities, proximity to employment, and so on, but income disparities between different districts in the same metropolitan area have reached disquieting levels in the United States. Researchers have shown that incomes in the richest 10 percent of census tracts in US metro areas are almost five times those of the poorest 10 percent, with the great majority of rich households living in new suburban locations and the poorest living in older neighborhoods in central cities and inner suburbs. Other research has shown that the spatial heterogeneity of income increases with the amount of sprawl in a metro area.[41] Suburban sprawl, it seems, has filtered people by income and magnified social inequity. On the one hand, it has allowed the rich to segregate themselves into clusters where they can escape the social problems and high taxes of inner cities, whereas on the other hand, it has concentrated the urban poor in distressed central cities. With falling real incomes since the early 2000s and a shrinking middle class, poverty is spreading geographically to older suburbs and beyond.[42] These spatial patterns are present in virtually every metropolitan area in the United States.

The concentration of extreme poverty in specific neighborhoods creates a web of cause-and-effect factors that lock people into unemployment and other opportunity-denying structures. Education is traditionally seen as an avenue out of poverty, but in US cities, the structure of the public system actually contributes to growing disparities between rich and poor. Public schools in poorer areas are disadvantaged by inadequate funding. Because almost half of public school education is financed from local sources, poorer districts will have less money to spend per student than tonier areas. Generally speaking, state funding does not close the gap. As a result, per pupil expenditures undertaken by states and localities in 2004–2005 were $938 higher in the quartile of all schools with the least poverty compared with the quartile of schools with highest proportion of low-income students.[43] Low-poverty schools are located overwhelmingly in suburban precincts, whereas high-poverty schools are predominantly found in central cities. Because per pupil funding is an important factor in student success, it is no stretch to conclude that local-source funding undermines the basic principles of equality of opportunity. Poor schools in the central city also contribute to the flight of middle-class families to the suburbs, where better public schools are available. This dynamic helps drive the income-filtering process and locks the metropolitan region into a system of communities differentiated by life opportunities.

Another equity issue relates to so-called job sprawl, or the tendency for employers to decamp to the suburban fringe where highways, highly educated workers, and well-off consumers have been increasingly found for many decades. Although the relocation of jobs to suburban areas benefits employers, who can offer lower wages to workers with shorter commutes, it harms those left behind in central areas who can no longer find suitable jobs close to home and who must lengthen their commute, take lower-paying work, or move into the ranks of the unemployed.[44] This spatial mismatch between location of residence and suitable work appears to affect African Americans most directly, undermining their access to job opportunities.[45]

This concern is related to another equity issue that has received increasing attention in recent years. Low-density, car-dependent development means that residents must have access to cars to be full economic citizens. Several studies have identified the lack of access to automobiles as a major disadvantage to poorer citizens in US metropolitan areas, essentially depriving them of opportunities to compete fairly in the job or educational market and improve their life conditions.[46] Smart growth policies have been shown to improve life opportunities of disadvantaged people by making low-cost transportation more accessible and effective.[47]

In some important ways, this spatial pattern of differential opportunity is reinforced by the pattern of public spending, patterns that are increasingly being questioned on an equity basis. The chief culprit is the federal deduction on home-mortgage interest payments, which are claimed disproportionately by suburbanites. In 2008, as opposed to the very visible $16 billion spent by the federal government to subsidize low-income renters, primarily in central cities, this hidden expenditure to home owners amounted to almost $100 billion. Other hidden subsidies to suburbanites take the form of highway-construction spending, low fuel taxes (compared with other rich nations), widespread provision of free parking, and spending on highway patrol areas.[48] This pattern of public subsidies not only contributes to growing inequality between urbanites and suburbanites, but it lowers the cost of living or doing business on the urban fringe relative to the core city and therefore contributes to sprawl itself.[49]

Signs of Change

The drivers of change summarized above have been emerging for some years, but today they are converging into a "perfect storm" blowing the winds of change through US cities. Real changes are taking place that herald a transformation in our urban environments. Many new development and transportation trends are emerging, such as new mixed-use and mixed-density neighborhoods around transit stops, redevelopment of grayfields into walkable communities, retrofits of commercial arteries, suburban infill projects, and proliferation of "complete streets" designed to accommodate different options. Two recent trends—the revival of downtowns and the move away from auto travel—are good indicators of the many changes taking place.

Revival of Downtowns

A strong city center is essential to a livable city. Healthy and vibrant downtowns provide a focus for the region's transit infrastructure, have walkable commercial and mixed-use districts, have a strong employment base, and can create a unique identity that provides the city region with a shared sense of purpose. In short, downtowns are convivial.

As sprawl took hold in the immediate post–World War II period, downtowns in the US declined. Downtown office and retail buildings were old and neglected after years of war and depression, and they could not compete with new retail and office districts built at the periphery close to the burgeoning highway network. As incomes rose and private cars became more widespread, downtown residents were attracted to cleaner, more spacious, peaceful, and leafy precincts rising out of farmland at the city's edge. As downtown property taxes receipts declined, city administrations cut services, and a vicious cycle of decline set in.

Believing that the best way to revitalize their centers was to make them car-friendly, city leaders pressed for freeways to be built around or even through the downtown, tore down trees to widen streets, demolished older buildings for surface parking, and transformed two-way street grids into one-way networks to enhance traffic flow. Unfortunately, these policies only worked to destroy the advantages that had kept downtowns healthy in previous eras: their compact, contiguous, densely built urban fabric. As it turned out, these desperate measures only served to accelerate the decline of downtowns and promote the suburbanization process.

Despite prognostications of their ultimate decline, many inner cities in Northern America are on the rebound. One after the other, cities are reshaping themselves by removing urban expressways, linking downtowns to regional subcenters via good-quality transit, and filling in parking lots with mixed-use buildings. Other signs of downtown revival include restoring natural features and creating major new parks, redeveloping disused port areas and brownfield sites, and expanding universities. Finally, many formerly derelict shopping streets in residential neighborhoods in or near downtowns are being revitalized through investments in street enhancements, better infrastructure, and policing.[50]

These efforts appear to be getting traction, as revealed by studies in the United States that reflect a growing population base in many US downtowns. A 2009 study of forty-four selected downtowns in the United States showed that between 2000 and 2007, 86 percent of the sample was showing increases in population. As author Eugénie Birch concluded:

> *The new paradigm for downtown (dense, walkable, mixed use with a heavy component of housing) is quite established in many of the nation's cities. While this downtown still has considerable commercial activity, its employment base is more diverse, with jobs in anchor institutions (universities; hospitals; and entertainment including arts, culture, and sports) rising as a proportion of the total. The residential component has become significant and is shaping the demand for neighborhood-serving retail, schools, and open space.*[51]

Of course, not every downtown is being transformed, but revival is clearly visible in cities that not too long ago had been given up for dead. Rust Belt cities like Buffalo, Pittsburgh, and Cleveland are attracting new residents by developing their waterfronts and fostering new areas of industry, especially centered around the growing health and medical research arena. Downtown revival stories are not limited to large urban centers, either. Even smaller cities like Wichita, Kansas, and small towns like Woodstock, Georgia, are on the rebound.

The demographic trends noted above are helping spur new growth in downtown areas. With more leisure time, some empty nesters are choosing to downsize their housing for the convenience of living in a downtown condominium. Added to them are young professionals in their twenties and thirties who have yet to start families and are frequently in the market for low-maintenance, urbane housing convenient to work and amenities. Those downtowns that have managed to preserve historic buildings and unique landscapes also offer a niche market for those seeking a "sense of place."[52] These emerging trends are signs of a long-term process that Alan Ehrenhalt calls the "great inversion," with middle- and upper-income families returning to city centers while suburbs become the refuges for minority populations, newly arrived immigrants, and lower-income households.[53]

Clearly, the forces of decentralization are still operating, and low-density development on the urban fringe marches on. It is also clear, however, that new forces are emerging that have a reconcentrating effect on downtowns. If these patterns continue, and there is no reason to doubt they will, we can expect downtown areas to continue their resurgence. The stronger downtown gets, the more likely it is that the surrounding central-city neighborhoods will strengthen as well. The overall results could be an enhanced interest in city living and a spurt in the demand for new housing in older parts of the city. A virtuous cycle could be set in motion.

Changing Travel Behavior

The United States once led the world in public transit use. At the turn of the twentieth century, grid-style street systems, rapid population growth, and a booming economy fostered a streetcar revolution that swept across the country. By 1920, Americans living in cities were averaging more than 250 transit trips per year, mainly on the nation's 40,000 miles of electric railway. Hundreds of US cities were served by privately operated streetcar lines, often put in place by developers who were also creating new subdivisions linked by electric trams to employment centers.[54] In the decades that followed, however, increasing prosperity, low fuel and vehicle costs, and increased investment in paved roads brought the private automobile to dominance. During the first twenty years following World War II, the nation's population grew 35 percent, but automobile registrations soared 180 percent in the same period, from 26 million in 1945 to 72 million in 1965.[55]

Until just a few years ago, the automobile continued its unchallenged rise, and the number of miles driven annually on US roads steadily increased. At the turn of the twenty-first century, though, something unprecedented occurred: motor vehicle travel demand growth slowed and then stopped. The number of vehicle miles traveled per capita peaked in 2004 at just over 10,000 and declined gradually after that until it reached a low of 9,409 in 2013, a reduction of about 7 percent. The figure rose slightly in 2014, but is still below the per capita driving level that pertained in 1997.

Of course, it is impossible to predict how these trends will play out in the coming years, but it should not escape our notice that the 2004–2013 per capita reduction in driving continued much longer than the longest previous periods of contraction on record, those associated with the oil crisis of the 1970s and the stagflation of the early 1980s. That this prolonged decline preceded the Great Recession and continued (with the exception of 2014) beyond it may suggest a permanent new trend toward lower levels of vehicle use.

A long-term trend is also detectable in the figures related to public transit ridership. In 2014, there were 10.8 billion transit trips taken in the United States, the highest absolute number in fifty-eight years. From 1995 to 2014, transit ridership increased by 39 percent, almost double the population growth, which was 21 percent. Over the same period, the growth of vehicle miles traveled was only 25 percent. The growth in rail-based trips—especially light rail—has accounted for much of the increase and now makes up 46 percent of total boardings, up from 34 percent in 1996.[56]

Of course, many factors could be involved in shaping changes in travel patterns like those mentioned. Some of the more evident ones are dramatic shifts in fuel prices, the effect of transit investments, the growth in telecommuting, and the gradual retirement of the baby boom generation. Working behind these objective trends, however, might be changing attitudes toward travel, especially among young people. As Richard Florida observes in an article in

Atlantic: "Younger people today—in fact, people of all ages—no longer see the car as a necessary expense or a source of personal freedom. In fact, it is increasingly just the opposite: not owning a car and not owning a house are seen by more and more as a path to greater flexibility, choice, and personal autonomy."[57]

Changing Housing Demand

Fueled by demographic change, concerns over quality of life, and changing transportation preferences, there has been a growing interest in more compact, walkable communities with a mix of land uses and realistic alternatives to the car as a transportation mode. Researchers have documented this shift in consumer demand through a variety of methods and concluded that demand for more livable and sustainable communities is outstripping the supply of such environments.

Housing prices are a handy proxy for the demand for housing in different locations in metropolitan areas. Studies have shown that consumers are willing to pay more for housing located in areas that are walkable, of higher density, and have a mix of uses and access to jobs and amenities such as transit. For example, Charles Tu and Mark Eppli found that, after controlling for other housing characteristics, buyers paid 4.1 to 14.9 percent more for housing in new urbanist developments.[58] Proximity to high-capacity transit has been shown to increase property values, a phenomenon known as the "transit premium." The Center for Transit Oriented Development examined a range of studies to determine the effect of transit investments on real estate values and found that transit premiums ranged from a few percent to more than 150 percent.[59]

The price premiums for transit- and walk-friendly locations may reflect a mismatch between the supply and demand for such neighborhoods. Emerging survey evidence points to a structural shift in real estate preference, away from far-flung, low-density suburbs toward mixed-use, compact, amenity-rich, transit-accessible neighborhoods or walkable places. For example, a survey conducted for the National Association of Realtors (NAR) in 2004 found that 46 percent of all Americans want to live within walking distance of public transit.[60] A more recent survey for NAR showed that the figure is growing, with 58 percent of home buyers surveyed preferred mixed-use neighborhoods where one can easily walk to stores and other businesses. Only 40 percent selected a community with housing only, where residents need to drive to get to businesses.[61]

Conclusion

There is no doubt that the ground is shifting under our cities. A demographic swing is in motion that is giving rise to a new class of home seekers demanding more urban environments and a wider range of travel options. More attention is being paid to public health issues, and awareness of how health outcomes depend on the built form of our cities is growing. The new emphasis on the quality of urban places is driven by rising expectations of a mobile workforce and the competitive strategies of urban regions. The financial limitations and risks of the sprawl model are becoming more apparent as our infrastructure debt climbs. The increasingly visible effects of climate change and the destructive effects of growing social inequality are two bellwethers that we must change how we are designing and building our cities.

As these forces continue to build interest in and shift demand toward smarter, walkable, human-scale neighborhoods, it will become increasingly important to remove the barriers that are preventing the supply of such built landscapes from catching up. At the local level, there are many standards, regulations, and subsidies that have given rise to exclusive low-density suburban communities over the decades since World War II. Large-lot, single-use zoning and expansive parking requirements are increasingly seen as obsolete rules that are preventing the emergence of new development patterns adapted to changing consumer preferences. Increasing attention is also being paid to the many local subsidies that have fueled low-density fringe growth, such as the failure to charge developers the full cost of the infrastructure needed to support new communities.

At the regional level, the absence of effective agricultural land conservation and growth management strategies has given all but free reign to developers to buy cheap rural land and convert it into low-density suburban landscapes, serviced mostly at public expense. The lack of coordination between land use and transportation in metropolitan regions has also hampered the emergence of more transit-friendly neighborhoods and sustained the default view that we need more roads to resolve congestion problems. The implications of this approach for the mounting problems associated with maintaining and upgrading the nation's infrastructure have helped unleash a much-needed national discussion about the best way to manage regional growth and plan major investments in our regional transportation systems. The paucity of regional tax-sharing arrangements has contributed to the historic decay of our inner cities.

At the state level, the lack of oversight afforded municipal planning has created a regulatory void that has led to widespread leapfrog, car-dependent development and limited the application of good planning principles in the design of new communities. State funding arrangements for public schools have created vastly different educational experiences depending on where children's parents live and contributed to both middle-class flight to the suburbs and poverty concentrations in inner cities. Federal mortgage rules that discourage investment in mixed-use, compact developments are also being questioned, as are subsidies to sprawl such as the tax break on mortgage interest payments by home owners.

The good news is that changes to the urban landscape are already under way. Downtowns are being revived in places across the country, and travel choices are moving in a more sustainable direction, with a new emphasis on transit, walking, and biking. Although unsustainable development continues in many areas, hundreds of projects that are planned or already built are beginning the slow transformation of our cities from the auto-dependent, single-use environments that characterize sprawled growth to the complete and walkable, human-scale communities that characterize smart growth. Whether they are downtown redevelopments, new greenfield villages, commercial corridor retrofits, mixed-use neighborhoods on former grayfield sites, or new streetcar-based suburbs, these projects demonstrate new models for reforming our practices of city building and stand as guideposts to the future.

As Simmons Buntin put it in his book *Unsprawl: Remixing Spaces as Places*: "Good public realm as a setting for human interaction, convenient amenities in proximity, and harmony with the natural settings were once considered elemental requirements of city-making, but they have been obliterated by the madness of building wasteful sprawl for half a century. But these principles have now been recovered."[62] Let us continue by providing some insights on how this gathering recovery can be strengthened using lessons from the Canadian experience.

Chapter 2

Canada and the United States: Similar yet Different

Despite Canada's proximity and its importance to the United States economy, ignorance of things Canadian is almost legendary in the United States, and one cannot assume that an American[1] reader will bring much if any background knowledge about Canada to this book. This chapter, then, offers a short overview of some of the historical, social, and political factors that distinguish Canada from the United States and that are likely to have a bearing, however indirect at times, on the two countries' different urban and suburban development outcomes.

Canada and the United States are, it must be recognized, very much alike in many ways. They share the greater part of the North American continent along with a frontier story stemming from the gradual extension of largely white settlement throughout the continent. They both see themselves, albeit perhaps inaccurately, as "new" countries for whom immigration is not only a reality but an important part of the national identity; it is not insignificant that people as different as John F. Kennedy and Mitt Romney have dubbed the United States a "nation of immigrants."[2] Both the United States and Canada have long traditions of democratic government grounded in long-lasting and resilient federal systems. They share, although with some exceptions, largely similar economic conditions, labor force characteristics, and demographic trends. Finally, they share a common language, with the obvious exception of Francophone Canada, and a largely common living standard and lifestyle. In contrast to Europe or Mexico, where the cultural differences with the United States are immediately apparent to the naked eye, Canada and the United States *look* alike, with Tim Horton's taking the place of Dunkin' Donuts and Esso in place of Exxon. Rural and suburban landscapes and housing types tend to be similar, and cities in Canada and the United States share a largely common vocabulary of structures, streets, and open spaces. These overwhelming yet superficial similarities tend to mask arguably even more important differences.

History and Identity: Who Are the Canadians?

Although both the United States and Canada came into being as a result of European colonization over the same era, two particular features distinguish Canada's history from that of the United States. One is the emergence of Canada as a nation divided between English and French speakers, a split that was not just one of language but one of religion and culture as well. The other is the different manner in which the two countries emerged from colonial status into nationhood.

Canada did not have a revolution. In contrast to the United States, one writer has described Canada as having been "created by another foreign power for reasons of its own that had little to do with Canadian desires and nothing to do with natural geographical boundaries."[3] That is perhaps something of an exaggeration. Although the creation of the Dominion of Canada (as it was known) in 1867 reflected some desire for self-rule on the part of many Canadians, it was equally driven by fear of potential United States northward expansion. As John A. MacDonald, who became Canada's first prime minister, said in the debates leading up to confederation: "The occasion of war [by the United States] with Great Britain has again and again arisen.... We cannot say but that the two nations may drift into a war as other nations have done before. It would then be too late when war had commenced to think of measures for strengthening ourselves, or to begin negotiations for a union with the sister provinces."[4]

This sentiment was not paranoid thinking. The United States had invaded Canada, albeit unsuccessfully, in both the Revolutionary War and the War of 1812, and the Oregon Territory dispute of the 1840s had led to vociferous saber-rattling on the American side. With the United States having become a major military power as a result of its mobilization for the Civil War, it was only reasonable for Canadians to fear their southern neighbor.

Thus, in contrast to the United States, where a coherent national identity that had already begun to emerge before the country itself was then forged in the course of war and revolution, Canada came to being more or less as an act of *realpolitik* rather than as a reflection of a polity that had already come into being. As a result, what it means to be Canadian was for generations—and to some may still be—a question or a work in progress.

Reflecting the nation's origins as well as its geographic position as a small country (in population if not in land mass) attached to a far larger one, Canadian identity has from the beginning been seen as much as a form of opposition to the dominant superpower to the south as something intrinsic to the Canadian nation. As early as 1839, Lord Durham, then Governor-General of British North America, wrote, "If we wish to prevent the extension of this [US] influence, it can only be done by raising up for the North American colonist some nationality of his own; by elevating these small and unimportant communities into a society having some objects of a national importance; and by thus giving their inhabitants a country which they will be unwilling to see absorbed even into one more powerful."[5] More recently, sociologist Seymour Martin Lipset has written that "Canadians have tended to define themselves not in terms of their own national history and traditions but by reference to what they are *not*: Americans. Canadians are the world's oldest and most continuing un-Americans."[6]

One particularly notable feature is the difference in what might be called the "frontier myth" in the two countries. The settling of the West is a central element in the history and identity of both countries, yet the central features of the story that grew up around it are very different. The US story is one of individualism, of the lone gunslinger, either as outlaw (Butch Cassidy) or as defender of the powerless (Shane); the Canadian story is one of law and order, and the hero is not the frontiersman, but the Mountie, sent by the national government to ensure what one Canadian historian has described as "a non-American type of development in the prairie west."[7] One can speculate that it is in this early iconic glorification of the law that the roots of the Canadian preference for law and order—and the sharply lower levels of violence, crime, and gun ownership in Canada—can be found.

Canadian identity, whatever precisely that means, remains in flux. As the *Canadian Encyclopedia* puts it, "From the beginning, the question of Canadian identity—of what it means to be a Canadian, what moral, political or spiritual positions it entails—has been a vexed one, so much so that some place the very act of posing it at the center of that identity."[8] Much of the uncertainty that resonates in this statement reflects the many fault lines running through Canadian society, all of which affect not only identity, but governance and social policy as well. Indeed, in many respects, the project of building a national identity in Canada is not so much developing an identity in itself, but finding one that transcends—and that can coexist with—the many separate identities that already exist within the Canadian nation or, in the alternative, accepting that Canadian identity is the sum of those separate identities. As Canadian historian G. R. Cook wrote on the occasion of Canada's centenary commemorations in 1967: "Perhaps instead of constantly deploring our lack of identity we should attempt to understand and explain the regional, ethnic and class identities that we do have. It might just be that it is in these limited identities that 'Canadianism' is found, and that except for our over-heated nationalist intellectuals Canadians find this situation quite satisfactory."[9]

That may or may not be the case, and the form that a distinct Canadian identity might take is still unclear. What appeared to emerge during the years after World War II was what one might call a *de minimus* identity, as Canadian social commentator Rudyard Griffiths noted: "Through the 1950s, '60s, '70s, and into the 1980s, there were three pillars of Canadian identity: peacekeeping, healthcare and the threat of Quebec separation."[10] In a similar vein, the winning entry of a 1972 CBC radio competition for a Canadian equivalent to the phrase "as American as apple pie" was not "as Canadian as hockey" or maple syrup, but rather "as Canadian as possible, under the circumstances."[11]

Canadian fault lines are linguistic, racial and ethnic, cultural, and geographic. The most powerful one, of course, is the continued divide between Anglophone and Francophone Canada, a divide that led to a violent separatist movement in the 1960s, the rise of the independence-minded Parti Quebecois in the 1970s, and the sovereignty referenda of 1980 and 1995. Passions have arguably cooled considerably since then, and secession is a far less central political topic today. As one recent writer commented, citing surveys that show a sharp drop in support for Quebec sovereignty among younger adults: "The mortal enemy of the sovereignty movement isn't the Liberal Party of Quebec, the Trudeau family, the federal government, Quebec's immigrant population or any of the other central casting nightmares conjured up by the sovereignist movement over the years. No, the real enemy is the march of time."[12] Although

Francophone and Anglophone Canada may have found a stable form of coexistence within a single overarching political entity, the two are and are likely to continue indefinitely to be two different nations in a fundamental cultural sense in a way that has no remote parallels in American regional cultural differences. The linguistic/cultural divide remains a central, perhaps *the* central, reality in Canadian society and politics.

It is not the only one. Regional divides, over and above those involving Quebec, resonate strongly. The three prairie provinces of Alberta, Manitoba, and Saskatchewan clearly share a cultural identity largely distinct from the rest of Canada, as does—albeit grounded in a very different cultural matrix—British Columbia. Atlantic Canada may also be seen as almost a separate country, particularly Newfoundland, which joined Canada only reluctantly in 1949 as a result of economic crisis. Newfoundland still retains a strong identity; although there is no significant separatist sentiment on the island, a 2003 survey found that 72 percent of residents responding identified first as Newfoundlanders, secondarily as Canadians.[13] There are strong regional divides in the United States, between north and south and between east and west. The concept of state's rights, moreover, forms an important part of the nation's political rhetoric, particularly on the political right. Even so, one would be hard-pressed to argue that state or regional identity trumps national identity in any part of the United States or that the idea of "going it alone" finds resonance except among small groups at the political extremes.

Ethnic and racial identity is another significant feature of the Canadian social landscape, although it takes very different forms than similar divides in the United States. Aboriginal populations play a much larger role in the Canadian scene than in the United States, if not directly, at least in terms of how they are seen as part of the national mosaic; the Aboriginal population of Canada[14] makes up 4.3 percent of the nation's population compared with less than 1 percent for the Native American population in the United States. The territory of Nunavut, which was created in 1999 and encompasses much of the Canadian North and most of the Canadian Arctic, is a self-governing Inuit polity within Canada in which Inuit languages share official status with English and French.

Although the idea of Canada as a nation of immigrants is yet another pillar of Canadian identity, the extent to which immigrants, who make up a large part of the Canadian population, have distinct identities separate from their identity as Canadians is also a matter of considerable attention and debate. Canadian author Andrew Cohen has noted, using the popular trope that Canada is a mosaic whereas the United States is a melting pot,[15] that "in Canada, there is an enduring belief that immigrants may remain largely unassimilated. If the melting pot demands conformity, the mosaic offers distinctiveness."[16] Although this point is seen by many Canadians as a point of pride, Cohen further argues that this rejection of assimilation as a goal "suggests that we all remain guests, old and new Canadians, reluctant to make an unconditional commitment to the country."[17]

Although Canadians are justifiably proud of their country—and according to some surveys are more satisfied with their lives than the residents of almost any other country[18]—when compared with the extent to which large numbers of Americans seem almost obsessed with what it means to be an "American" and with its associations of power and wealth and with such elusive yet pervasive ideas as American exceptionalism, Canadian identity may not unreasonably be characterized as "identity lite." It would be hard to imagine someone from the United States saying about his or her country, as Griffiths recently did, that "things like equalization[19] and healthcare bind a lot of the country together, and there's something sort of noble in the idea that if you live in Newfoundland you should have roughly the same access to healthcare and quality of life as someone living in, say, downtown Calgary."[20] An approach to identity in which social policy is as central as these examples suggest is likely to carry with it implications for how people relate to their institutions of government, how they perceive their relations with others within their society, and how that affects their behavior.

These differences can arise from historical factors such as the manner in which Canada came into being or particular features of its economic or political systems. They can also be the product of different underlying cultural and civic values, or some combination of the two, which clearly affect one another in a circular, reinforcing relationship. Not surprisingly, people have not only speculated but have studied in some detail the questions of whether "Canadian values" are in fact different from "American values" in ways that explain the differences between the two societies and whether those values are converging, diverging, or neither. As a result, a considerable body of survey data that attempts to answer these questions has accumulated. Values in this respect can be both those that deal with one's personal beliefs and behavior and those that deal with external matters, such as one's position on the role of the state or on specific issues of policy.

Dutch social psychologist Geert Hofstede has defined culture as "the collective programming of the mind which distinguishes the members of one group or category of people from another," and has developed measures along various dimensions to show the cultural differences between societies.[21] His analysis of the United States, Canada, and Quebec shows significant differences on many of those dimensions. The differences, where they are present, are clearly matters of degree rather than the fundamental differences one might expect—and indeed would find—between the United States and a society operating within a fundamentally different cultural matrix, such as Japan or one in the Arab world. Still, there are significant differences in a number of areas, particularly in the areas of individualism and masculinity, where Canadians express a clear preference for a less—but not radically less—individualistic society and one in which caring is given greater weight than competition. Notably, the divergence between Quebec and the United States on these two values is even greater, suggesting that the value difference between Americans and Canadian Anglophones is less than a direct comparison between US and Canadian respondents would suggest, although still present.

Other researchers have identified significant differences with respect to specific values; Canadians tend to be less religious[22] and less patriarchal; Canadian researcher Michael Adams found in 2000 that 49 percent of Americans but only 18 percent of Canadians agree with the statement "the father of the family must be master in his own house."[23] Similarly, a 1996 survey found that 49 percent of Americans owned guns, whereas the same was true of only 22 percent of Canadians,[24] and a 2011 study found that the United States had 89 civilian firearms per 100 residents, whereas Canada had 31 firearms per 100 residents.[25] Finally, an analysis of the World Values Survey for the period 1981–2000 confirms many of these differences, but it also argues that the value trajectories of the two countries parallel each other—generally in the direction established by Canada rather than the other way around—but are not converging.[26] Interestingly, for the value "income should be made more equal," although both societies are showing increased support for the proposition, the shift was much greater in Canada, leading to a substantial divergence by 2000 on a value where there was little divergence in 1981. The values shown in figure 2-1 translate readily into attitudes toward public policy, not only with respect to universal health care but also with respect to gun control, abortion rights, and—two areas where the United States is catching up with Canada—gay marriage and decriminalization of marijuana. This finding is consistent with other research that shows that Canadians are generally less individualistic, more supportive of government programs, have greater trust in government institutions, and are more supportive of government spending than Americans.[27]

The role that these value differences play with respect to the differences between United States and Canadian policies governing land use, growth, and urban development can only be a matter for speculation; we would suggest, though, that it is a not insignificant one. Although they do not *define* those differences, which stem from legal, institutional, and other factors, they form a critical part of the underpinning for them. That said, however, it should be stressed that cultural and social values are far from immutable. It can be argued

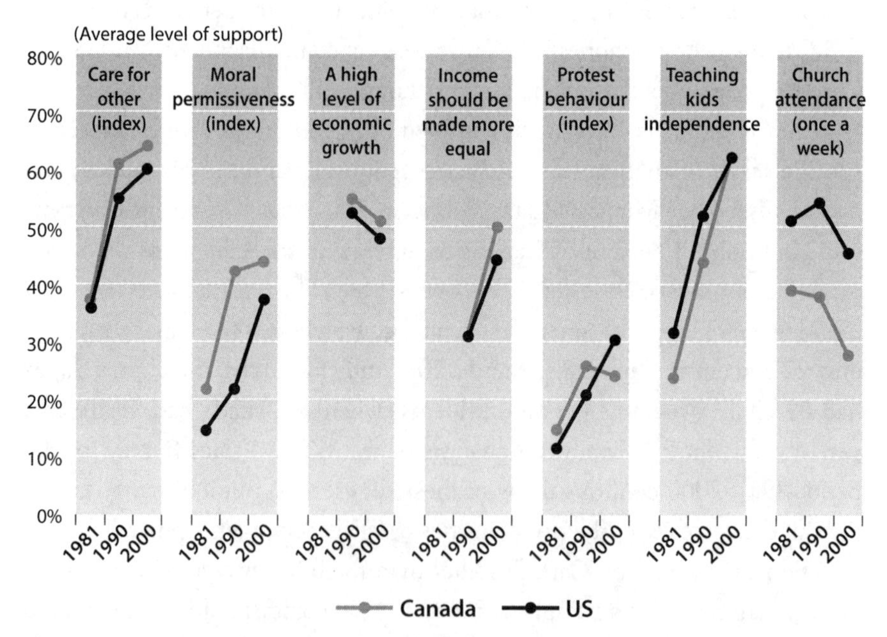

FIGURE 2-1. Changing values in the United States and Canada, 1981–2000
Source: World Values Survey, described and analyzed in Boucher 2014.

that the high point of a redistributional social policy as Canadian identity was as far back as the 1970s and that it may be less central to the Canadian polity today than it may have been at that time. Certainly, the Conservative administration of Prime Minister Stephen Harper, in power as we write, does not appear to share many of what may be considered traditional Canadian values or some of the historical perspectives on what it means to be Canadian, and it appears to be eager to narrow rather than extend the historic divergence between Canada and the United States in this respect as well as in others.

Geography, Society, and Economy

Canada is a very large country with a relatively small population. Although slightly larger in area than the United States, it has only 11 percent of its population. At the same time, however, it is a highly urbanized country, with the great majority of its population living within a narrow band of little more than 100 miles from the United States border[28] and the great majority of its land area unpopulated (figure 2-2). According to the World Bank, 81 percent of the population in both Canada and the United States lived in urban areas in 2013.[29]

Canadian and United States cities share a common history. They have been formed over nearly four centuries, beginning with settlements on or near the Atlantic driven by European immigration and gradually moving across the continent as industry, agriculture, and mining drove both population movement and the creation of urban centers. By the twentieth century, rural populations had begun to diminish, and the populations of both countries were concentrating increasingly in cities and their surrounding suburbs, forming metropolitan areas. Since the end of World War II, the population in metropolitan areas has grown from 56 to 84 percent in the United States and from 45 to 71 percent in Canada.[30] The most significant difference in urban growth between the two countries emerged after the war, as large-scale internal migration in the United States led to explosive growth in the Sun Belt cities of the South and West, a phenomenon that Canada was geographically incapable of replicating.

Table 2-1 presents population data for the ten largest metropolitan areas in each country and their growth between 1951 and 2011 (for Canada) or 1950 and 2010 (for the United States). In Canada, 35 percent of the population is concentrated in its three largest metropolitan areas—Toronto, Montreal, and

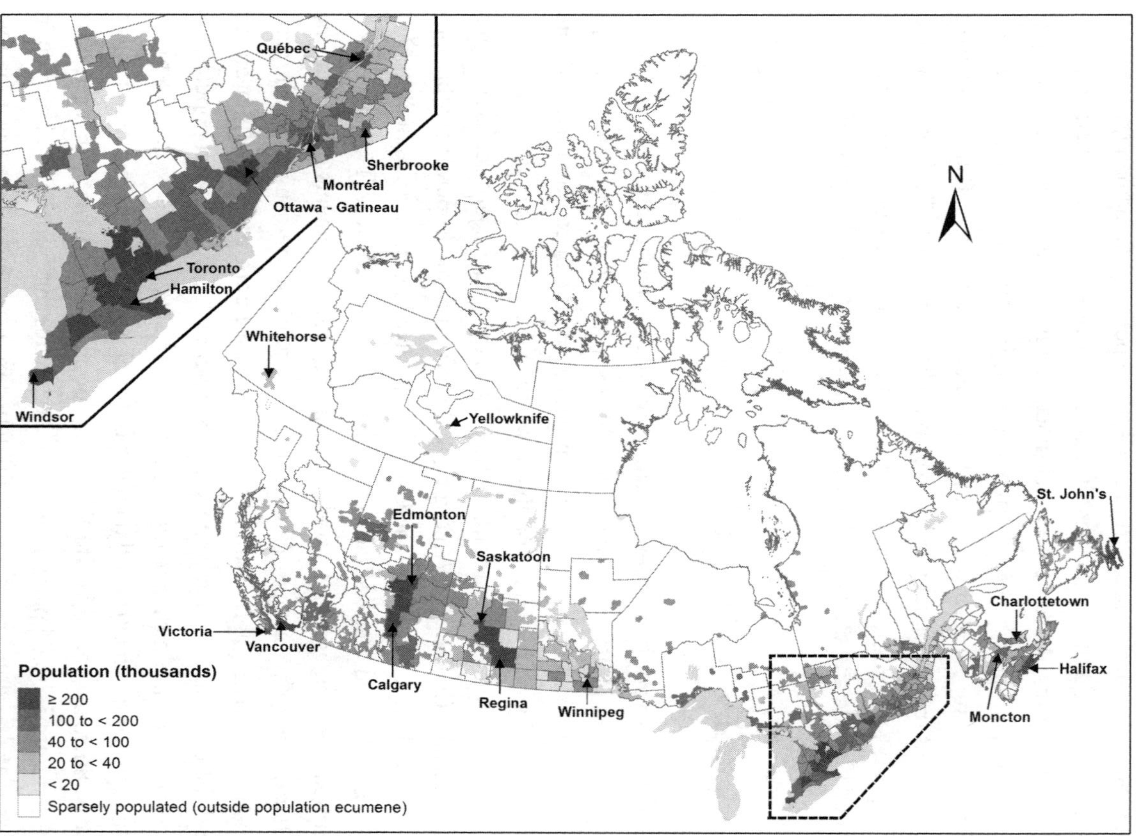

FIGURE 2-2. Spatial distribution of Canada's population
Source: Statistics Canada, 2011 National Household Survey.

TABLE 2-1.
Ten largest metropolitan areas in Canada and the United States

Canadian Metros	1951	2011	Growth 1951–2011	US Metros	1950	2010	Growth 1950–2010
Toronto	1,262	5,583	342%	New York	13,589	19,567	44%
Montreal	1,539	3,824	148%	Los Angeles	4,368	12,829	194%
Vancouver	586	2,313	295%	Chicago	5,761	9,461	64%
Ottawa-Gatineau	312	1,236	296%	Dallas–Fort Worth	1,262	6,426	409%
Calgary	142	1,215	755%	Philadelphia	3,973	5,965	50%
Edmonton	194	1,160	498%	Houston	1,083	5,920	447%
Quebec	289	766	166%	Washington, DC	1,721	5,636	228%
Winnipeg	357	730	104%	Miami	694	5,565	702%
Hamilton	282	721	156%	Atlanta	1,091	5,287	385%
London	168	457	173%	Boston	3,187	4,552	43%

Source: Compiled from Statistics Canada and US Census, from demographia.com.

Vancouver—compared with 14 percent of the US population being concentrated in New York, Los Angeles, and Chicago. That difference reflects both the more polycentric nature of the United States, unsurprising in light of its greater population and the greater extent of its economically productive area, and the effect of Sun Belt expansion on its population distribution. The only Canadian metropolitan area large enough that it would be one of the ten largest metros if it were in the United States is Toronto.

In both the United States and Canada, the lion's share of metropolitan growth has taken place in the suburbs; between 1950 and 2000, the central-city share of metropolitan population in the United States dropped from 59 to 34 percent. Although the same phenomenon was taking place in Canada, it is obscured statistically by the widespread expansion of urban boundaries that has taken place through annexation or amalgamation. If, however, one looks at Canadian central cities in their 1951 borders, as we do in chapter 4, one can readily see that Canadian suburbanization has followed trends not much different from those of its southern neighbor. In short, with respect to both urban concentration and subsequent suburban deconcentration within metropolitan areas, Canada and the United States have followed largely parallel tracks.

Both the United States and Canada have advanced economies that follow generally capitalist models. The features of those economies are largely similar, although Canada's far smaller population means that its economy is much smaller, with major implications for the US-Canada trade relationship. Both countries are among the world's wealthiest and most productive nations, although most indices place the United States slightly ahead of Canada. The Canadian economy suffered through the Great Recession as well, but it was largely spared the effects of the collapse of the housing bubble and the subprime mortgage sector, principally because of more conservative lending practices as well as more effective banking regulation. At least partly as a result, home-ownership rates have remained high in Canada, and as of 2013, 69 percent of Canadian households were home owners compared with 65 percent in the United States. Because Canada does not allow home owners to deduct mortgage interest from their income taxes, this disparity adds support to the argument that the home-mortgage interest deduction actually has little or nothing to do with a nation's home-ownership rate.

Part of this difference may also be attributable to the income distribution in Canada being significantly less unbalanced than in the United States. In 2010, the Gini coefficient, a widely used measure of income inequality, for Canada was 33.7, which was slightly above the average for all member countries of the Organization for Economic Cooperation and Development but substantially below that of the United States (41.1).[31] Most of that difference, however, is attributable to the substantially more robust Canadian social safety net rather than fundamental differences in wage structure. It may be relevant, however, that as of 2013, 30 percent of all Canadian workers were members of trade unions compared with 11 percent in the United States.[32] Similarly, 9 percent of Canadians were below the low-income cut-off as of 2011, a measure comparable to the US poverty level, whereas 16 percent of Americans were below that level.[33] Income inequality is rising in Canada, but it still has a long way to go before it is comparable to US levels.[34]

The Canadian economy, similar to that of the United States, is heavily driven by services. Just the same, it is noteworthy that a significantly larger share of Canada's gross domestic product and its workforce are still based in manufacturing, with 28 percent in the goods-producing sector compared with 19 percent in the United States. Agriculture and natural resource extraction—mining, oil and gas, and logging—still play a greater role in the

Canadian economy than in the United States and accounted for about 58 percent of Canada's total exports in 2009. This difference reflects what may be considered a significant imbalance in the Canadian economy: although the majority of the natural resource extraction sector is oriented toward the export trade, the nation's manufacturing sector tends to be more limited to the domestic market and includes a large share of branches of foreign, largely US, firms.[35] Although the United States accounts for 73 percent of Canadian exports, Canada represents only 23 percent of US exports.

Despite these variations, however, the economies of Canada and the United States share far more common ground than they differ, and they are highly integrated with each other and with the global economy through international trade and cross-national company ownership. Differences in some important social characteristics of the two countries are somewhat more significant, however.

Both countries are ethnically diverse, but, as shown in table 2-2, their ethnic mix is quite different. Although racial/ethnic minority members in the United States are largely African-American and Latino, they are predominately Asian in Canada. Although Canada has a small but not insignificant black population, more than two-thirds of that population are immigrants from Africa or

TABLE 2-2.
Ethnic/racial distribution of population

	Canada Visible Minorities 2006 (%)	**United States Race/Ethnicity 2013 (%)**
Black/African American	2.5	12.3
Latino/Latin American	1.0	17.1
Asian	12.8	5.0
Other visible minority	0.5	—
Other race/two or more races	—	3.2
Not visible minority[a]	83.8	—
White/not Latino	—	62.4

Source: Compiled from Statistics Canada Census; 2013 American Community Survey.

Note: The manner in which Canada and the United States classify populations for statistical purposes, as can be seen from the table, is somewhat different; the definitions are nonetheless similar enough so that the comparisons, with some room for error, can reasonably be made.

[a]Roughly comparable with the US category "White/not Latino."

the Caribbean. By contrast, the overwhelming majority of African-Americans in the United States are descended from the men and women brought to the country as a result of the slave trade.

Both countries are important immigrant destinations, but since the 1950s Canada has steadily absorbed a significantly larger flow of immigrants relative to its total population than the United States, a pattern that continues to this day. That may surprise Americans, who are accustomed to the acrimonious debate about the number of immigrants being admitted to their country. In fact, in the first decade of this century, the number of immigrants admitted to Canada represented 6.6 percent of its current (2011) population, whereas those admitted to the United States made up 3.8 percent of their 2013 population,[36] and foreign-born residents make up 21 percent of Canada's population compared with 13 percent of that of the United States. Thus, although Canada is absorbing more new immigrants, as indicated in table 2-2, the United States is, and continues to be, a more racially mixed nation.

Canadian and US immigrants come from different places. Table 2-3 shows the place of origin of recent immigrants to both countries. Although the data do not permit one to pinpoint the *current* share of immigrants to the United States coming from Canada, it is worth noting that 2 percent of the foreign-born population in the United States is Canadian-born compared with 4 percent of the Canadian foreign-born population having originated in the United States.[37]

TABLE 2-3.
Place of origin of immigrants, Canada and the United States

Origin	Canada (2006–2011) (%)	United States (2000–2009) (%)
Asia	57	28
Europe	14	8
Africa	13	*n/a*[a]
Latin America and Caribbean	12	56
United States	4	—
Other areas	1	8

Source: Compiled from Statistics Canada 2011 National Household Survey and 2013 American Community Survey.

[a] Not applicable; included in "Other areas" category.

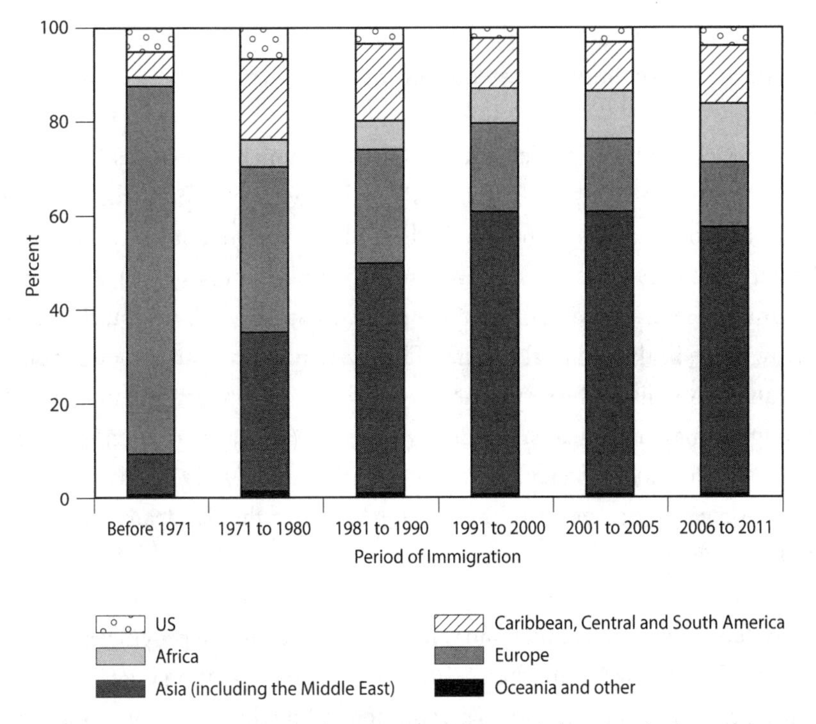

FIGURE 2-3. Region of birth of immigrants to Canada by period of immigration
Source: Statistics Canada, 2011 National Household Survey.

As shown in figure 2-3, the pattern of immigration to Canada has shifted significantly over time. Prior to 1971, immigration was predominately from Europe. During the 1970s and 1980s, as the Asian share grew, roughly 20 percent of immigrants were from the Caribbean and Latin America, predominately Afro-Caribbean in composition. Since the 1990s, more than half have been from Asia, with a growing share coming from Africa. The principal Asian countries from which immigrants have come to Canada in recent years have been the Philippines, China, and India.

Institutions, Politics, and Governance

Both the United States and Canada are long-standing federal representative democracies, in one case for well over 200 years and in the other for nearly 150 years. That similarity is not only significant; it is fundamental to understanding the reality of the two countries. There are nonetheless important differences both in the two countries' systems and the underlying principles governing them.

First, Canada has a parliamentary system, in which candidates are elected by districts to the House of Commons[38] and in which the party that wins the largest number of seats is then invited[39] to form the government. One effect of a parliamentary system, as distinct from the presidential system adopted by the United States, is that it offers more scope to third, and even fourth, parties to play a role in the political system. This system has had an important effect on Canadian political history. The left-wing New Democratic Party (NDP), although it may never lead a national government,[40] has played an important role in shaping social policy, particularly during years when no party held an absolute majority, as was the case during many of the Trudeau years in the 1970s.

Second, and arguably more important, the division of power within the Canadian federal system is much more heavily oriented to the provinces than the US system is oriented to the states. Ironically, such orientation is generally recognized to be the opposite of what was intended by the two countries' respective founders. In the United States, a constitution that initially was designed to give the new nation a weak central government and strong states has been gradually reframed and redefined over time to create a polity in which the national government clearly dominates the states. By contrast, the British North America Act, which created the Dominion of Canada in 1867,[41] contemplated a strong central government similar to that of the then highly centralized United Kingdom.[42] A long history of court decisions and legislative actions reinforced by the centripetal pressures inherent in the intense Canadian regional differences, however, has led to a gradual strengthening of the role of the provinces, to a point where the Canadian system has come to resemble far more closely the weak federal system initially envisaged by the founders for the United States than that intended by its own founders.

Moreover, the British North America Act enumerated certain explicit powers that were reserved for the provinces, which were restated in Canada's 1982 constitution. As these powers have been defined over time, provinces have retained or gained control over urban affairs, education, and natural resources, and the federal government has been constrained from acting directly in this area.[43] Thus, when it comes to matters of urban growth and development, the provinces are effectively sovereign. Further reinforcing the central role of the provinces in the Canadian polity is the subservient role of local government.

Neither in the United States nor Canada do local governments have constitutional status. Although the states are sovereign over local government in the United States, in practice that sovereignty has been widely and often severely curtailed by state constitutions, legislation, or custom. "Home rule," or the grant of relatively broad powers to local government to manage a wide variety of matters without direct state government action, is widespread in the United States.[44] Home rule powers are embodied in many state constitutions; a typical one is that of Montana, which provides that "a local government unit adopting a self-government charter may exercise any power not prohibited by this constitution, law, or charter."[45] Even in states that lack such constitutional language, de facto home rule may emerge from legislation or from the political culture and customs of the state. Thus, although the power of the state to adjust municipal boundaries, or compel two or more municipalities to consolidate, exists *in theory* in every state of the United States, it is exercised so rarely as to be all but unknown.

In Canada, neither constitutions, statutes, nor customs limit the control that the provinces have over their constituent municipalities. Indeed, as the cumulative outcome of a series of institutional and fiscal decisions and arrangements made over the past century, provinces have become clearly supreme. Provinces can and do modify municipal boundaries, consolidate municipalities into regional entities, create (and dissolve) regional transportation or planning entities, and create bodies such as the Ontario Municipal Board that can reverse local planning decisions. Indeed, although the 1998 amalgamation by the province of Ontario of six separate municipalities into a Toronto "megacity" was strongly opposed by all six municipalities, the provincial government persevered, and the amalgamation became law.

Despite such controversies, there is compelling evidence to suggest that the Canadian system has indeed played a significant role in fostering healthier cities and more compact urban form in Canadian metropolitan areas than in the United States. In chapter 3, we will begin to elucidate the nature of those differences.

Conclusion

This admittedly short survey of differences and similarities between the United States and Canada should make clear that, notwithstanding important similarities arising from similar economies and urbanization trends and a largely shared cultural heritage, the two countries are significantly different in many important respects. Moreover, although there may be some parallels in the trajectories currently being followed by the United States and Canada, whether in terms of values or more concrete matters such as economic inequality, there is little support for the proposition that the two countries are converging in these respects.

The relationship between the Canadian system of governance and certain outcomes with respect to urban form and development is arguably explicit, as will be discussed in later chapters, but we suggest that underlying values and practices, whether with respect to immigration, economic inequality, or something as fundamental as the value placed on "caring for others," translate into different urban outcomes. Although it is impossible to make these connections explicit, it is worth the reader's while to bear these differences in mind.

Chapter 3
Differences in Livability and Sustainability

Since the 1980s, the word *sustainability* has gradually entered our lexicon and captured our hopes for steering the world in a direction that is gentler on the environment, more socially just, and economically healthy. Of more recent vintage is the concept of *urban* sustainability, which refers to a city that minimizes the use of natural resources and production of wastes while providing a livable, socially inclusive milieu for its inhabitants with sufficient economic opportunities to meet their needs and aspirations for the future.

Every continent has its own brand of urbanization. In the United States, we see a particularly destructive form of sprawl that involves moving out in an ever-expanding circle of low-density housing, big-box stores, office parks, and franchise strips. This urban landscape almost demands the use of private vehicles to reach its dispersed corners. This process of sprawl clearly has plenty of winners; it would not have taken root if it were not producing benefits for a wide swath of interest groups and social strata. As many advocates of the status quo have pointed out, low-density suburbs meet many people's needs for privacy, contact with (albeit manicured) nature, and choice in terms of housing type and employment opportunities.

The problem—and it is a very big one—is that this type of urbanization is only possible in a society that can afford to commandeer an impressive proportion of the world's resources in the form of building materials, vehicles, consumption goods, energy, land, and water to build and operate what must be one of the most inefficient living and working arrangements ever invented. Not only do sprawl and car dependence consume massive resources, they produce vast quantities of wastes, including wasted energy (in the form of heat), air contaminants, greenhouse gases, and polluted water. They also have enormous effects on the livability of our cities in that they erode social solidarity by filtering people into different spatial zones based on race and income, concentrate poverty, undermine social mobility, and contribute to sedentary lifestyles and associated chronic diseases.

The global forces that produce sprawl operate both in the United States and Canada, and the symptoms of sprawl are becoming more evident everywhere. In this chapter, we track some indicators of those symptoms to see if they differ on the two sides of the border.[1] We divide the indicators into three categories: livability, environment, and resilience (a concept that includes both livability and environmental dimensions).[2] For each indicator, we will briefly discuss how it links to sprawl and see how metropolitan areas in each country perform.

Livability Indicators

With their lively downtowns that are relatively safe around the clock, good quality public schools, efficient transit systems, and walkable neighborhoods, Canadian cities are well known for their high quality of life. For good reason, Vancouver, Montreal, and Toronto—Canada's biggest cities—are considered by many to be among the most livable in the world. Although Canadians complain about the high cost of housing in some cities and the maddening traffic congestion in others, even these features can be seen as signs of success; these cities are places where people want to live and work, even if the costs are high.

Although livability is a widely sought-after feature of urban life, the exact meaning of the term is not easy to define. In general, livability includes dimensions such as the quality of the built and natural environments, economic prosperity, social stability and equity, adequate physical infrastructure, public health, safe streets, and educational opportunity. In other words, a livable city is one that has the social conditions necessary to support personal and community well-being. Many of the dimensions of livability can be shown to be linked to the degree of sprawl and car dependence that characterize urban areas. We present a few livability indicators here; we later explore these and other issues, such as crime and public education, in chapter 7.

Livability Indices

Several organizations track livability in cities around the world, each using its own set of metrics to create a unique livability index. Two such reporting initiatives—the Economist Intelligence Unit's Liveability Ranking and Mercer's Quality of Living Reports—are of particular interest because they include several US as well as Canadian cities. These assessments calculate the livability of cities around the world through a combination of subjective life-satisfaction surveys and objective determinants of quality of life.

The Economist Intelligence Unit's Liveability Ranking surveys 140 cities worldwide and ranks them using thirty qualitative and quantitative indicators across five broad categories: stability, health care, culture and environment, education, and infrastructure. Three of the four Canadian cities included in the index (Vancouver, Toronto, and Calgary) ranked in the top five worldwide, and the fourth (Montreal) was not far behind at number 16 in the world. US cities were ranked from number 34 (Washington, D.C.) to number 56 (New York) in the world.

Mercer's Quality of Living Survey ranks 221 cities worldwide on thirty-nine factors, including political, economic, environmental, personal safety, health, education, transportation, and other public-service factors. Of the five Canadian cities included in the index, all were ranked in the top 15 percent of the cities surveyed, with Vancouver, Toronto, Ottawa, and Montreal being the highest ranked of Northern American cities. San Francisco, at number five, was the only one of the seventeen US cities included in the index that outranked a Canadian city, sixth-place Calgary (figure 3-1).

The Economist and Mercer indices vary in terms of the indicators included and the weights given to various components. The overall pattern of results is clear, however: Canadian cities consistently score higher than US cities on international surveys of livability and are usually in the top tier of cities internationally.

Public Health

The relationship between urban sprawl and public health variables is complex, but there is increasing consensus that spread-out communities reduce our ev-

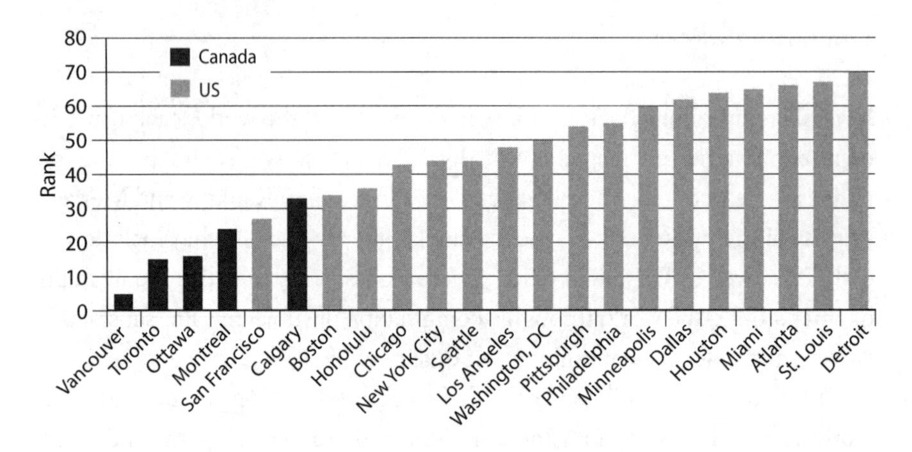

FIGURE 3-1. Mercer's 2015 quality of living survey, ranking US and Canadian cities
Source: Mercer's Quality of Living Survey, 2015.

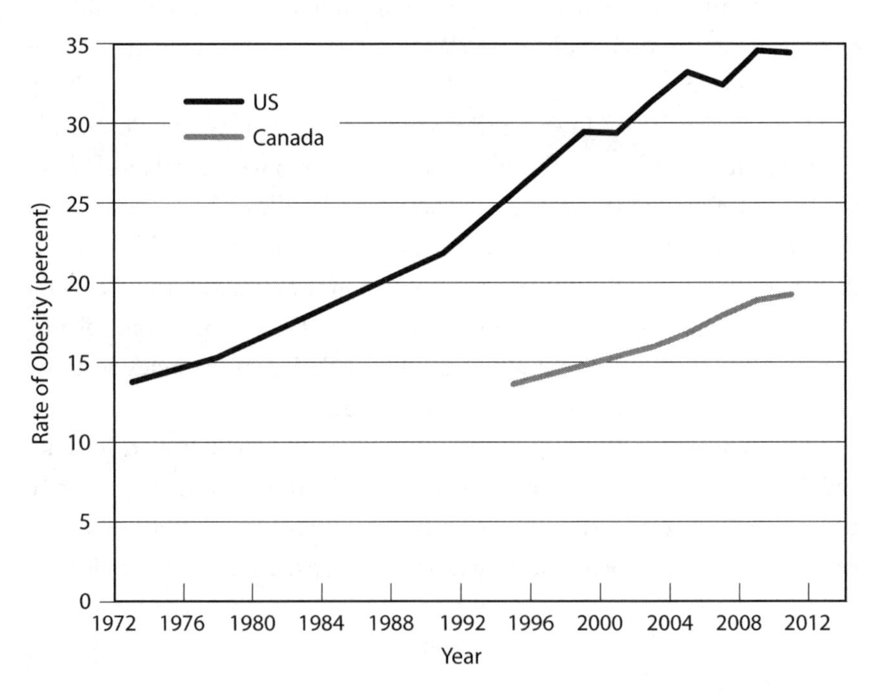

FIGURE 3-2. Obesity rates for Canada and the United States, percent of population, 1972–2012
Source: Organization for Economic Cooperation and Development, 2014, Obesity Update.

eryday level of physical activity by depriving residents of locations accessible on foot or bike and by enforcing the use of motorized means of transport. One indicator of this dynamic is the increasing rate of overweight and obese individuals in the United States and Canada, both high by international standards. As shown in figure 3-2, the long-term trends are the same in both countries, but the problem has achieved more alarming levels in the United States, where almost 35 percent of the population is considered obese compared with less than 20 percent in Canada.

Racial/Ethnic Enclaves

Apart from Aboriginal peoples, both the United States and Canada are nations of immigrants, a fact of great historical, economic, and cultural importance that we discuss in greater detail in chapter 7. The countries differ considerably, however, in the racial composition of their cities and the degree to which minorities are concentrated in urban enclaves. These realities have significant implications for the livability of urban areas.

Studies of ethnic and racial segregation in Canada have found that although certain groups are concentrated in specific neighborhoods, the levels of concentration for such groups were not as high as in US cities.[3] Moreover, the most concentrated ethnic and racial minorities—such as clusters of Chinese people in suburban Toronto and Vancouver—have above-average incomes and rates of home ownership. Levels of segregation in Canada for most minority groups have declined, particularly for the two minority groups (blacks and Aboriginals) that have some of the strongest associations with low levels of income. The one exception is the Toronto region, where segregation levels for these two groups as well as others have increased.[4]

A study comparing ten metropolitan areas in the United States with the same number of metros in Canada showed that spatial segregation between whites and non-European immigrant groups was lower in Canadian cities for five out of six immigrant groups (table 3-1).[5] The authors concluded that a greater degree of urban sprawl in the United States aggravated existing cleavages by multiplying municipal jurisdictions outside central cities where non-immigrants could barricade (figuratively speaking) themselves from immigrant

TABLE 3-1.
Residential segregation between whites and selected non-European immigrant groups, United States (2000) and Canada (2001)

Group	United States (%)	Canada (%)	Difference (%)
Iranians (US)/West Asians (Canada)	67.5	58.7	–8.8
Chinese	60.2	49.1	–11.1
Filipinos	56.6	49.1	–4.7
Latin Americans	55.4	69.5	+14.1
Caribbeans	55.3	47.7	–7.6
Asian Indians (US)/ South Asians (Canada)	52.0	50.5	–1.5

Source: Teixeira, Li, and Kobayashi 2011.

groups. The result is extremely uneven development between the central city, where a disproportionate number of poor and racial minorities live, and the suburbs, where most affluent whites live.[6]

Social Inequality

As just mentioned, sprawl is driven in part by affluent people escaping the problems of the central city, a trend that leaves the central city with fewer resources and more problems. Increasingly, this dynamic is affecting older suburbs as yesterday's leafy havens are in turn abandoned by the well-to-do for greener pastures farther from the central city.

A crude indicator of city-suburb social equity is the difference in aggregate socioeconomic characteristics between central-city and suburban residents. Table 3-2 summarizes income ratios for eighty-five US metro areas and eight Canadian ones. On average, the city-suburb household income gap is smaller in the Canadian cities; the average income in Canadian central cities is 89 percent of the metropolitan value versus 82 percent in the United States. Moreover, the US sample shows a much wider variation in income ratios, with many central cities experiencing disparities far beyond what is found in Canada. The largest gap in Canada was in Quebec City, where central-city residents earned 75 percent of what suburbanites made. In the US sample, the largest gap was

TABLE 3-2.
City-metro income ratios, selected metro areas, Canada (2001) and the United States (2000)

	Household Income			Population
	Central City	Metro	City-Metro Ratio	City-Metro Ratio
Canadian average	$45,556	$50,943	0.89	0.50
US average	$46,633	$57,485	0.82	0.33

Source: Taylor 2015.

in Hartford, where the central-city household income was only 47 percent of the metropolitan area as a whole.[7] We discuss this subject further in chapter 8.

The greater equality in Canadian metros is partially the result of many central cities having been expanded or consolidated to encompass more of the regional population, a process that is relatively rare in the United States. This difference is reflected in the population column of table 3-2, which shows that, on average, half the metropolitan population is included in the Canadian central cities, but only a third is captured in US central cities More consolidated regions tend to have smaller city-suburb income gaps because their central cities capture a greater share of well-off residents who would otherwise live in independent suburban jurisdictions. The smaller income gap also reflects that Canadian central cities have not been abandoned to the degree found in many US metropolitan areas and thus retain a greater population diversity in their urban cores.

Environmental Indicators

Despite its green international image, Canadians know that their country is no environmental paragon. A highly publicized report by the country's own Suzuki Foundation ranked Canada twenty-fourth out of twenty-five Western countries using twenty-eight environmental indicators; the only consolation is that it came in ahead of the United States, which ranked dead last.[8] Canada's national performance is compromised by the heavy reliance of its economy on resource extraction, especially fossil fuel production. For indicators not linked to energy, such as protecting natural areas, recycling, air pollution control, sewage treatment, and pesticide use, Canada was in the middle of the pack or better. When it comes to one-on-one comparisons with the United States, Canada's environmental credentials further improve, especially if we focus at the urban level where the effects of sprawl are most noticeable.

Sprawl leaves its footprint on the environment mainly through the conversion of rural land to urban uses, through the proliferation of large homes that require more resources to run (especially energy and water) and produce more waste, and by boosting unsustainable forms of urban travel, which in turn is reflected in higher levels of air pollution and greenhouse gas emissions. Sprawling development eats away productive agricultural lands, prime wildlife habitat, wetlands and aquatic systems, and other important components of local and regional ecosystems. In the United States, urban sprawl is estimated to have consumed about 4.2 million acres of prime or unique farmland between 1982 and 1992.[9] Unfortunately, we do not know of any consistent cross-national studies of land consumption so are unable to present comparative data here. Many of the other environmental dimensions of sprawl, however, can be compared based on relatively consistent metrics.

Green City Index

One source of data for anyone interested in comparing the United States and Canada on environmental performance is the *US and Canada Green City Index,* a comparison of twenty-seven large cities (twenty-two in the United States and five in Canada) using thirty-one indicators of urban greenness.[10] The index includes sixteen quantitative indicators of current environmental conditions in the sample cities; the remaining fifteen indicators relate to environmental policies and programs. The quantitative indicators—the ones of principal interest to us in the context of this chapter—cover issues such as carbon dioxide (CO_2) and air pollution, energy use, land use, transportation, water use, and waste.[11] The Canadian cities in the sample scored better than the sample US cities on average on twelve of sixteen indicators. A panel of experts was asked to weigh the quantitative indicators in terms of their importance to urban sustainability, and the cities were ranked accordingly. New York ranked

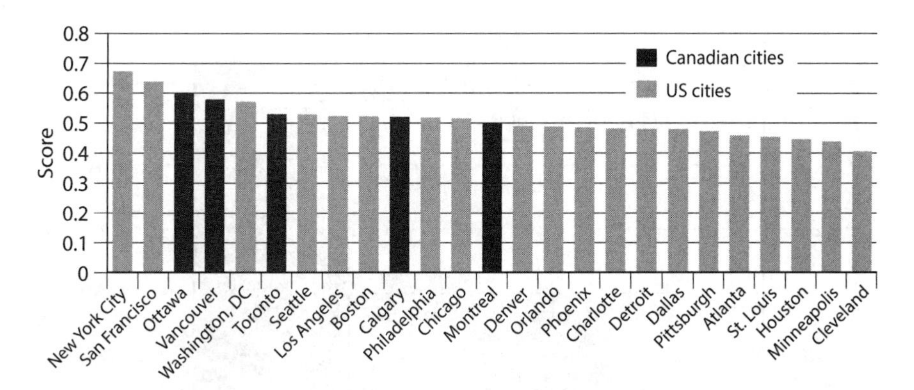

FIGURE 3-3. Weighted score on US and Canada Green City Index, 2011
Source: Compiled from Gokhan, Gumus, and Kucukvar 2015.

highest, with Cleveland the lowest. The Canadians cities on the whole did very well. Ottawa, Vancouver, Toronto, and Calgary scored in the top ten, and Montreal had a middling score (figure 3-3).[12]

Electricity and Water Consumption

Sprawled development means fewer multifamily buildings and more single-family homes on larger lots. Larger dwelling units require more energy to heat, cool, and run appliances. Sprawled development usually means more bathrooms, larger lawns, more swimming pools to fill in backyards, and more cars to wash in driveways. The differences between US and Canadian cities in terms of electricity and water use are significant, especially for electricity. As shown in figure 3-4, Canadian cities in the Green City Index sample used 31 percent less electricity and 13 percent less water than the US cities in the sample.

Greenhouse Gas Emissions and Air Pollution

One of the most telling indicators of urban sustainability is the quantity of greenhouse gases that a city emits. Because greenhouse gases such as CO_2 and methane arise from many urban processes, ranging from industrial processes to

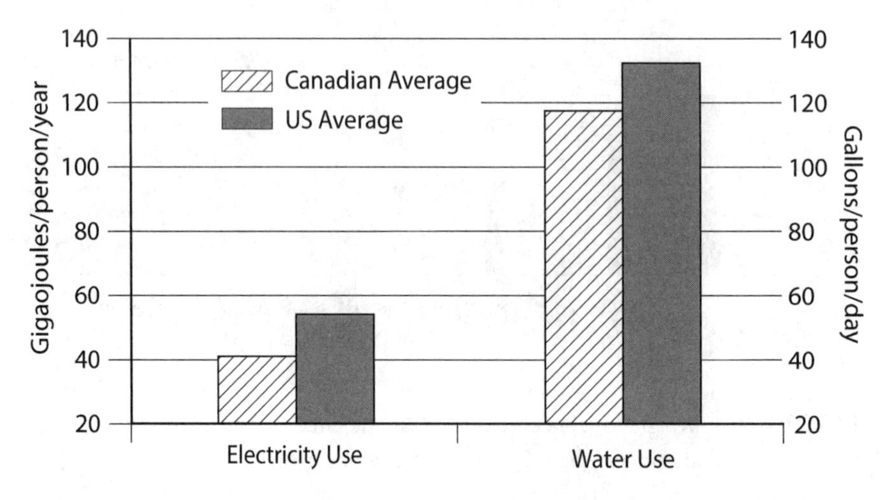

FIGURE 3-4. Electricity and water use, selected Canadian and US cities, 2009
Source: Economist Intelligence Unit 2011.

the decomposition of landfill waste, the quantity that a city emits is a good reflection of its overall burden on the planet's resources. Although each urban region's greenhouse gas emissions profile is unique, most show that a large share is due to issues linked to urban form, namely the use of energy in the heating and cooling of buildings and in urban transportation. Figure 3-5 shows average greenhouse gas emissions for the metropolitan areas covered in the Green City Index. Canadian urban regions emit just over half the CO_2 per capita as do their US counterparts: 9.4 metric tons per person compared with 16 metric tons for the US regions in the sample. Lower levels of economic productivity in Canadian cities and a greater amount of electricity generated from low-emission sources such as hydro and nuclear may account for some of the differences in per capita emissions, but the contrast between the two countries is large enough to conclude that differences in urban form and transportation factors play a role. That the biggest per capita emitter in Canada is Calgary comes as no surprise because it is a relatively low-density, car-oriented city and—as the capital of Canada's oil patch—has the lowest gas prices in Canada. Still, at 12.7 metric tons per person, Calgary's emissions are well below the average for the US regions in the sample. The lowest emitters of the twenty-seven cities in the sample were Ottawa and Vancouver, tied at about 7 metric tons per person. These

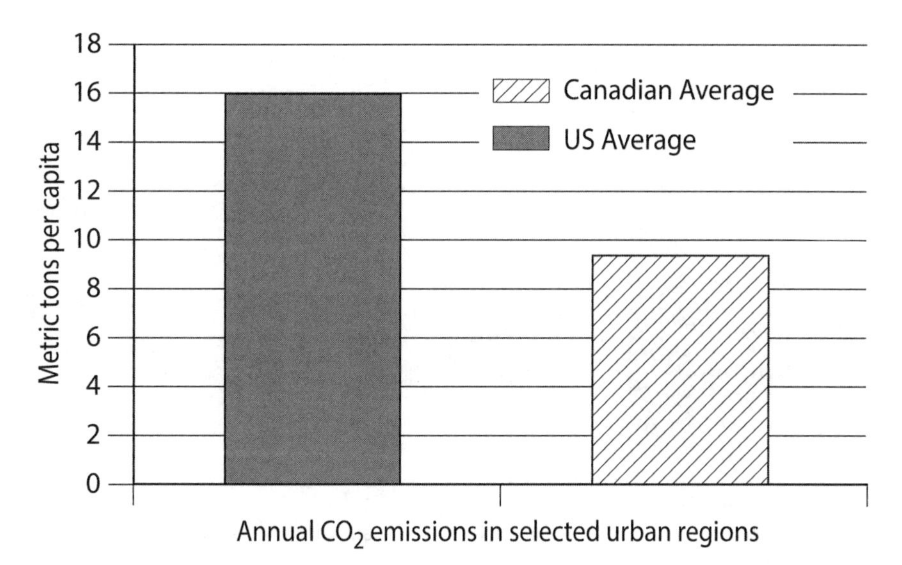

FIGURE 3-5. Annual CO_2 emissions in selected US and Canadian urban regions, metric tons per capita
Sources: Economist Intelligence Unit 2011; Kennedy et al. 2009.

cities have good-quality rapid transit and bus systems, relatively high densities, and extensive greenbelts that have helped stem sprawl. The biggest emitter in the United States was Cleveland (29 metric tons per capita), and the lowest was Los Angeles (8 metric tons per capita).

The Green City Index also measured annual emissions of three air pollutants; nitrogen oxides, particulate matter less than 10 microns in size (PM10), and sulfur dioxide. Like CO_2 emissions, these contaminants can be traced to multiple processes, but vehicle emissions are considered a major source within cities. Concentrations are typically much higher along busy roadways than in other parts of a community, increasing health risks to road users and nearby residents. The Green City Index showed that Canadian cities have lower emissions of all three pollutants (figure 3-6). The difference in PM10 (which was judged by the expert panel mentioned above as being of far greater importance to urban sustainability than the other two air contaminants) is especially striking: on average, Canadian cities had half the particulate emissions of their US counterparts.

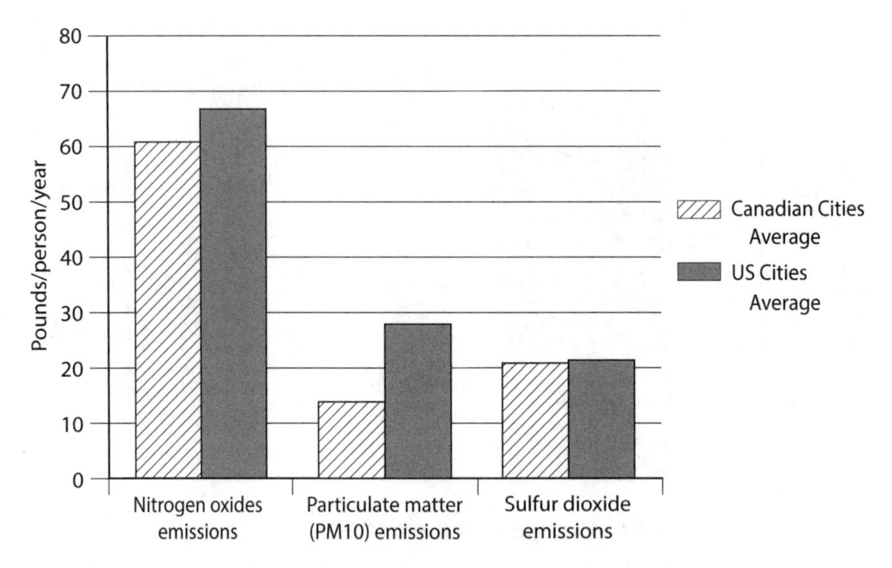

FIGURE 3-6. Annual air pollutant emissions in selected US and Canadian cities
Source: Economist Intelligence Unit 2011.

Travel Mode Shares

A good indicator of travel patterns in an urban region is commuting mode share, the breakdown of total trips during the morning commute by private vehicles, transit, walking, and biking. The data show that Canadians have long lived in cities that are less automobile-dependent and more convenient for transit and walking than US cities. In *The Myth of the North American City*, Goldberg and Mercer reported that in the mid-1970s, 85 percent of US commuters drove to work compared with only two-thirds of Canadians (figure 3-7). Equally significant was the difference in public transit usage: whereas one-fourth of Canadian commuters used public transportation, only one-eighth of US commuters did. Walking to work was also much more prevalent among Canadian versus American urbanites (8 vs. 5 percent).

After updating Goldberg and Mercer's numbers to 2010–2011, we found that the Canadian lead on transit, walking, and biking had been maintained into the twenty-first century. Although driving has increased its share in both countries, there is still a significant cross-border difference: 74 percent of Canadians choose autos compared with 90 percent for US commuters. Although no Canadian city

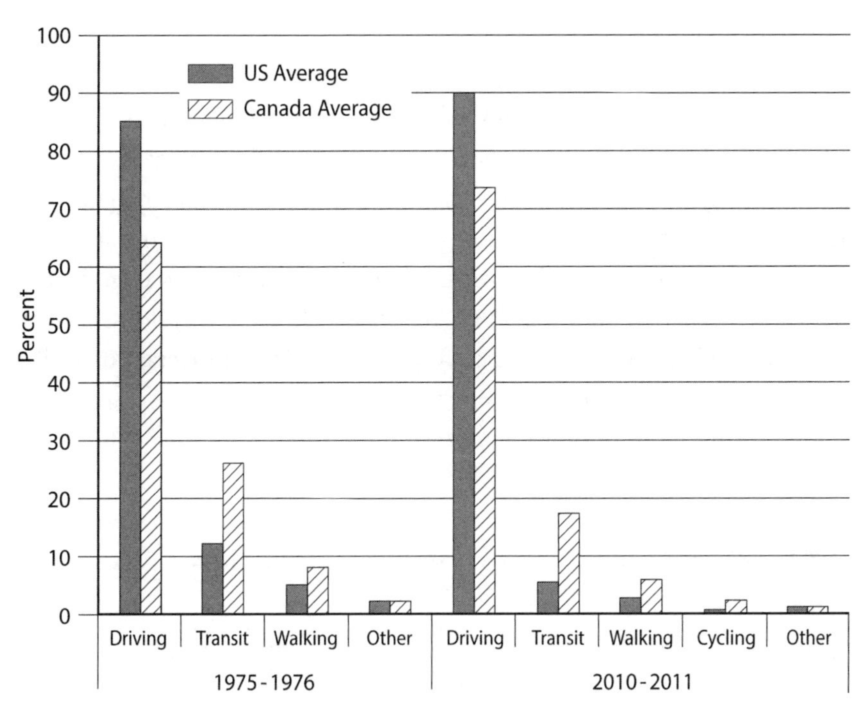

FIGURE 3-7. Mode of transportation for journey to work in the largest metro areas in the United States (1975, 2010) and Canada (1976, 2011)
Sources: Goldberg and Mercer 1986; Statistics Canada, 2011 National Household Survey; US Census Bureau, 2010 American Community Survey.

is as transit-oriented as New York, even Canada's more car-dependent urban areas are far less auto-oriented than most of their US counterparts. For example, Edmonton is a relatively small and—by Canadian standards—sprawling city on the prairies, but other than New York, Chicago, Boston, and Washington, DC, it has a higher transit share than any other US metropolitan area.

One might assume that the colder climate in Canada would deter bicycling, but in fact the reverse is true: cycling levels are considerably higher in Canadian cities. In 2011, the proportion of work trips by bike was more than three times higher in Canada than in the United States in 2010. With almost 6 percent of commuters traveling on two wheels, hilly Victoria, British Columbia, had the highest bike share of any city in the sample, which suggests that terrain can as be as easily overcome as cool weather. Miami, a flat city, has only a 0.3 percent cycling share.

Distance to Work

Cities built at higher density and with a greater mix of land uses generally have shorter trip distances. As shown in figure 3-8, the average length of a work trip in Canadian metropolitan areas is only about half that in the United States, across all population size categories. For both countries, the larger the population size, the longer the average distance of the journey to work. The increased trip length, however, is far greater for US cities than for Canadian cities. For the United States, the work trip lengthens from an average of 6 miles in the smallest population size category to 10 miles in the largest size category. In Canada, the work trip lengthens as well, but only from 3.4 to 5.3 miles.

Motor Vehicle Ownership and Gasoline Consumption

Given that Canadians tend to favor driving less than Americans, we should not be surprised to find that Canadians are also distinctive in terms of their vehicle ownership patterns. Figure 3-9 shows historic trends in per capita

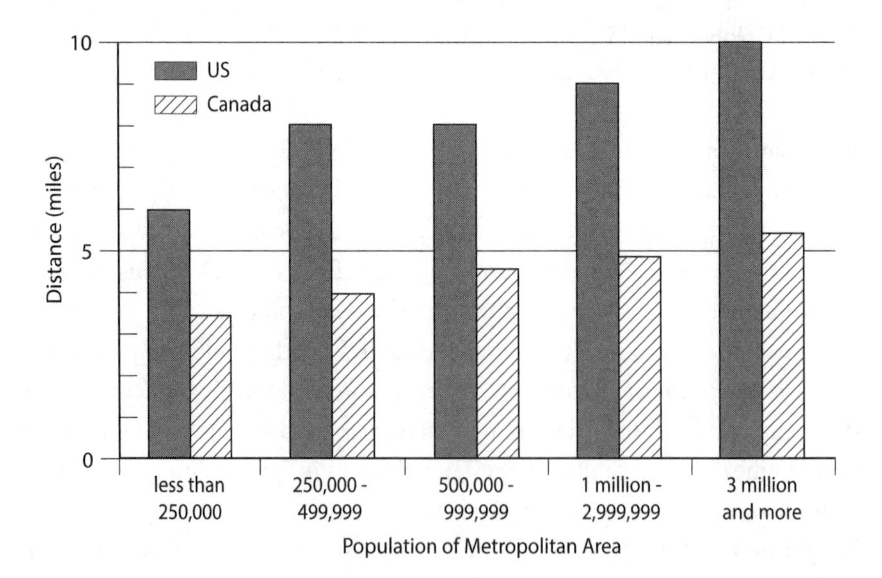

FIGURE 3-8. Distance to work, US (2000) and Canadian (2001) metro areas
Source: Pucher and Buehler 2006.

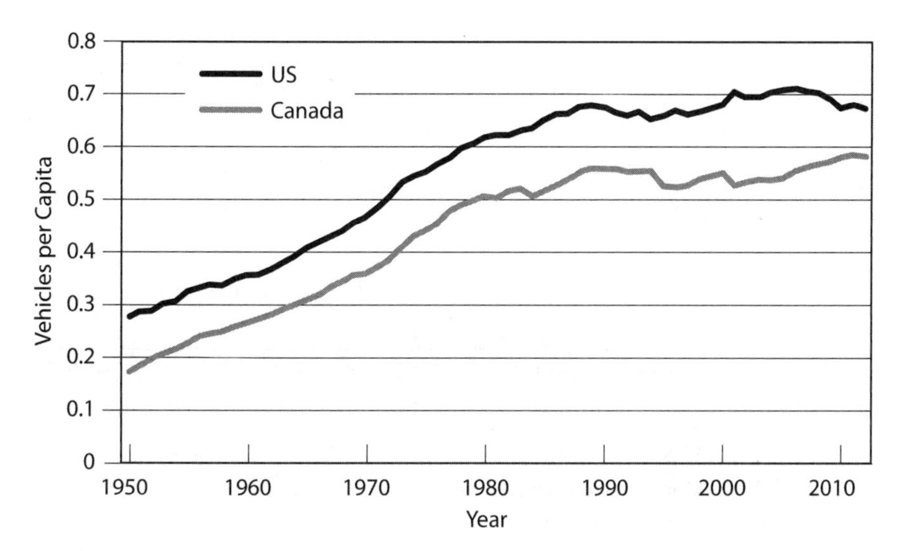

FIGURE 3-9. Automobiles and light trucks per capita in the United States and Canada, 1950–2012
Sources: Statistics Canada, Canadian Motor Vehicle Traffic Collision Statistics; US Federal Highway Administration, Annual Highway Statistics Series; some data courtesy of Paul Schimek.

automobile and light truck ownership in the two countries. The Canadian trend parallels the US trend, but at a consistently lower level. The number of vehicles per capita on the road increased dramatically in both countries after 1950 and then more or less leveled off at the end of the 1980s. The gap between the two countries has widened and narrowed over time, but in 2012 it was about the same as it was in 1950.

Per capita gasoline consumption in the two countries rose after 1950 as the number of vehicles per capita climbed steadily (figure 3-10). During the 1970s and early 1980s, world political events led to a rapid rise in gas prices (discussed at greater length in chapter 6), which precipitated a fall in gasoline consumption. After that, tighter efficiency standards for vehicles sold in both countries tended to keep per capita consumption from rising. The enduring difference between the two countries since then can be attributed to the higher price of gasoline in Canada; the tendency for Canadians to favor smaller, more efficient vehicles; and the lower level of car use that is associated with less sprawling cities.

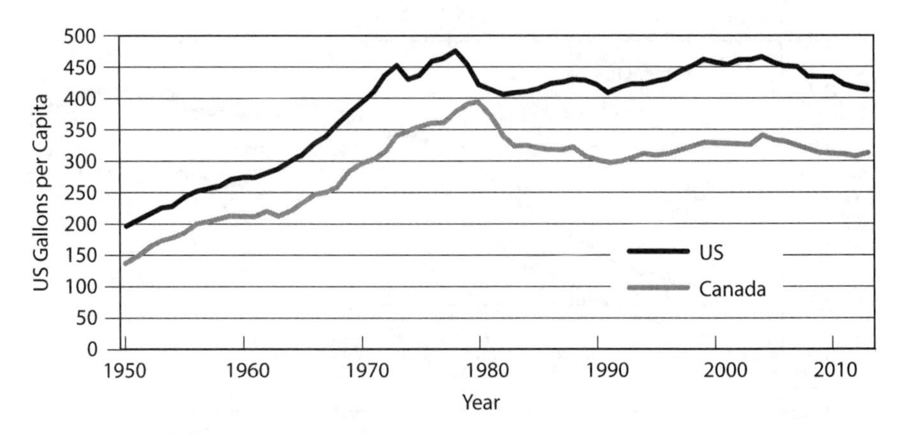

FIGURE 3-10. Gasoline consumption in the United States and Canada, 1950–2013
Sources: CANSIM tables, Human Activity and the Environment: Annual Statistics; US Federal Highway Administration Statistics.

A final measure that highlights differences in vehicle use between the two countries is the annual miles traveled totaled across all the vehicles on the road. If Canadians tend to own fewer cars, have a shorter commute, and tend to use noncar modes more than Americans, we would expect to find a lower level of miles driven per capita. Figure 3-11 tracks these numbers from 1968 to 2014. Some of the Canadian data are missing, but the overall trend is clear: Canadians drive significantly less than their American counterparts, and the difference appears to be gaining in magnitude. In the years around the turn of the twenty-first century, Canadians were driving only 60 percent the number of miles as Americans.

Transit Ridership

Another indicator of how environmentally friendly Canadian cities are compared with their US counterparts can be found in the fare box of transit vehicles. Rides taken on transit vehicles have a much smaller effect on energy consumption and greenhouse gas emissions than trips taken in private vehicles. One of the main determinants of transit ridership in any metro area is the degree of sprawl because less dense cities have a hard time providing effective transit services.

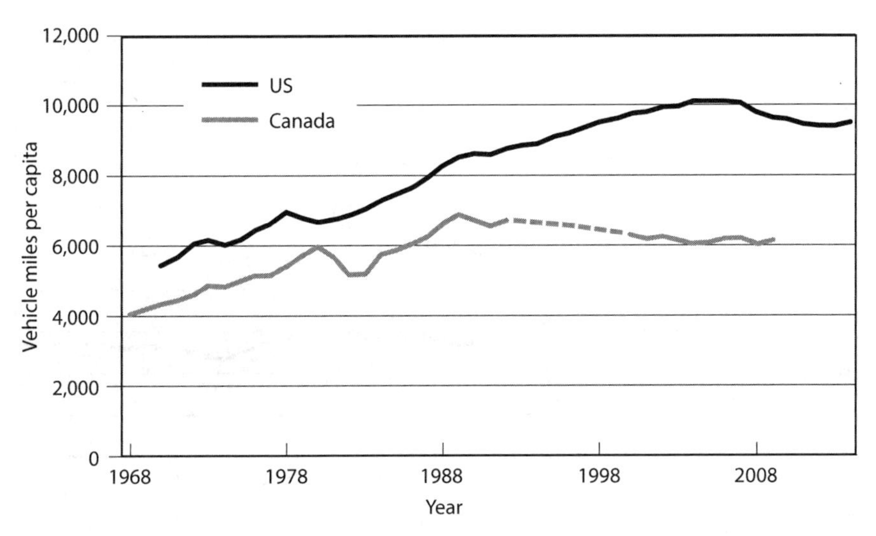

Note: Missing data marked with a dashed line.

FIGURE 3-11. Vehicle miles per capita in the United States and Canada, 1968–2014
Sources: Statistics Canada, Canadian Vehicle Survey, 2000–2009; US Federal Highway Administration, Annual Highway Statistics Series; some data courtesy of Paul Schimek.

Per capita transit use has been consistently higher in Canada than in the United States over the entire post–World War II period (figure 3-12). During the 1950s, transit use in both nations was coming down from wartime peaks. In the 1960s, the two countries diverged as per capita trips bounced back in Canada due to major investments in buses and rapid transit systems, but continued to fall in the United States until it stabilized at low levels in the 1970s. Since then, per capita transit use has fluctuated in Canada in response to waves of investment in transit services, whereas in the United States, per capita ridership has remained stubbornly low despite major investment in commuter rail and other transit improvements. Over the decades, the gap between the two countries has gradually widened, with transit use per person in Canada now at twice the US level. Clearly, higher transit use goes along with lower car use, two features of Canadian cities that are at least partially attributable to a lower level of sprawl in that country. We discuss this issue further in chapter 6.

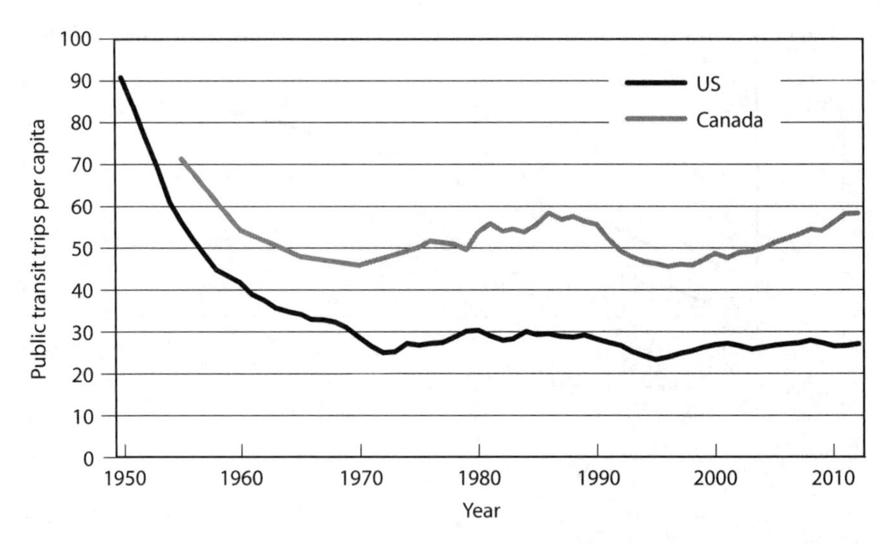

FIGURE 3-12. Public transit trips per capita, United States and Canada, 1950–2013
Source: American Public Transportation Association Fact Book, 2014.

Resilience

Urban resilience refers to the ability of a city to adapt to change while sustaining its essential functions as a center of human habitation, production, and cultural development. The concept is coming to the fore of urban sustainability discussions because of the increasing level of disruption being experienced by cities from the effects of climate change, globalization of trade, mass migration from rural areas, aging of the population, and depletion of essential resources such as oil. Cities can be assessed based on their vulnerability to these disruptive forces and their capacity to adapt to them. Vulnerability arises from exposure to disaster risks, inadequate infrastructure, environmental stress from pollution and urban sprawl, and social disparities that weaken communities. Capacity to adapt reflects the city's ability to plan for and manage change.

The resilience of major cities around the world has been measured by Grosvenor, a multinational real estate and investment firm, using a variety of indicators that reflect different levels of vulnerability and adaptability to disruption.[13] Of the fifty cities surveyed, three are in Canada and eleven in the United

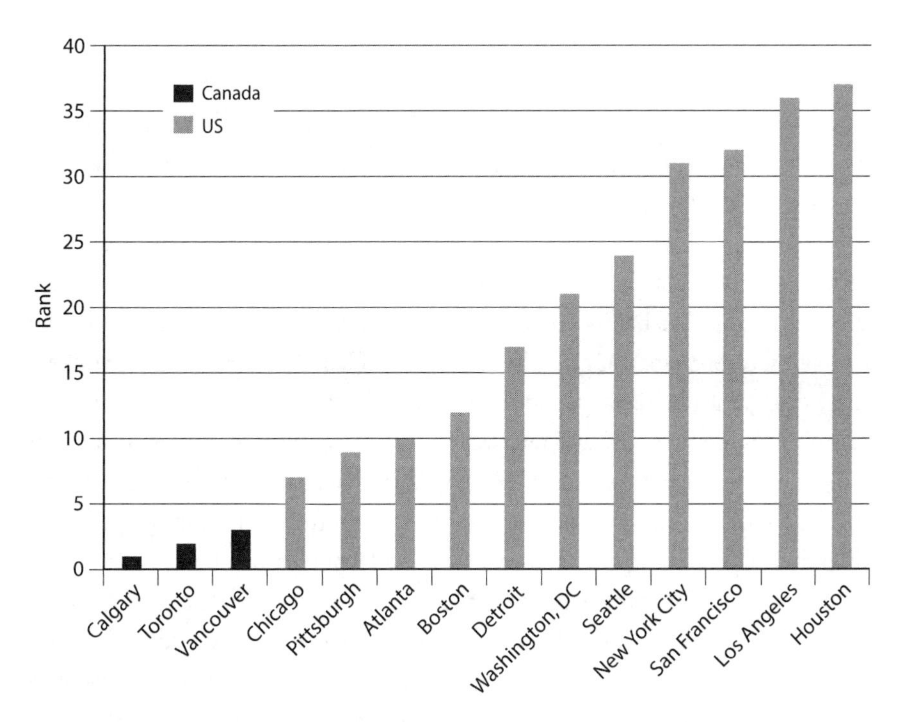

FIGURE 3-13. Grosvenor vulnerability rankings, US and Canadian cities
Source: Grosvenor 2014.

States. Based on the strength of their governing institutions, both Canadian and US cities performed well in terms of their adaptive capacity. Canadian cities outperformed US cities on the vulnerability index, however, which is of greater interest to us in the context of this chapter because it combines environmental and liveability measures. Grosvenor found that Canadian cities are less vulnerable to disruption because they are less stressed environmentally and socially, in part due to lower levels of urban sprawl. In contrast, inequality in US cities has led to social tension, and urban sprawl has contributed to the overconsumption of resources. Figure 3-13 captures the disparity between Canadian and US city scores on the vulnerability index, with the three Canadian cities scoring at the top of the index and the US cities scoring between seven and thirty-seven out of fifty global cities. The three Canadian cities in the study also scored the highest in the world on the resilience index, which is calculated by combining the vulnerability and adaptability indices.

Conclusion

Urban sustainability is a multidimensional reality, and no set of indicators can possibly hope to capture all the nuances involved in such an all-encompassing concept. Certainly, we make no claim to have done so. We chose indicators that are most relevant to our purpose: to explore whether Canadian cities show fewer signs of the malaise associated with urban sprawl. Of course, other indicators might have been added to fill out the details,[14] but we believe that the overall picture is clear. On the livability side, Canadian cities are more likely to be economically, racially, and ethnically inclusive, with inhabitants who are less prone to being obese and overweight. On the environment side, Canadian cities are more resource efficient, better balanced in terms of their transportation options, use less energy, and emit fewer greenhouse gases and air pollution than their US counterparts. However these advantages are achieved by Canadian cities, it is not at the expense of their livability; as we've shown, Canadian cities are consistently ranked among the most livable in the world, well above most US cities. Finally, the impressive resilience of Canadian cities in the face of expected pressures from climate changes, resource depletion, and demographic changes suggests that their advantages will not dissipate over time.

That cities in the two countries differ on average seems to be beyond doubt given the evidence before us, but how different is different? We know that when placed on a wider international scale, Canada and the United States tend to clump together on many sustainability indicators, but that does not mean that differences at the urban level across the two countries are trivial. We have shown that on a number of important indicators, urban conditions in both countries are strikingly different, conditions that are widely recognized as important not only to the fate of those who live in our cities but to the fate of the planet itself. The Canadian model might hold some secrets as to how cities in the United States could steer themselves toward a more sustainable future while preserving their essential qualities as prosperous, growing, stable, and democratic cities of the New World.

To unravel these secrets, however, we must first understand how the livability and environmental differences between the two countries arose. If variations in sustainability are linked to patterns of urban development, as implied in this chapter, just how different is Canada's urban form from that of the United States, and what forces wrought those differences? We turn to the first part of that question in the next chapter.

Chapter 4
Differences in Urban Form and Transportation Systems

Urban form, or the shape that a settlement takes on the ground, is key to describing and understanding any urban region. A region's urban form not only underlays its character and distinctiveness but has a major effect on its sustainability and livability. Its overall size and density, the location of major employment and residential areas, and the fine-grained texture of land uses in the different precincts of the region all affect how the region works. Urban form influences how urbanites get from place to place, their access to work and other life opportunities, the availability of housing of different types, and the overall prosperity of the region.

Our particular focus in this book is the interaction between urban form and transportation systems, essentially two sides of the same coin. Urban form affects transport activity by influencing how far people have to travel to get to their destinations and their choice of travel mode, whether that be by car, transit, walking, or cycling. Dense, mixed-use urban areas make a variety of transportation modes feasible, whereas low-density, single-purpose suburban areas tend to have fewer travel options.

The urban form of cities in both the United States and Canada has evolved substantially since World War II. What were generally dense, monocentric regions centered on downtowns with radiating transit services have evolved into dispersed, polycentric matrices that are generally far more reliant on automobiles. Central cities, once the industrial, transportation, cultural, and retail hubs of their regions, have been immersed in a much more sprawling urban region with a loose structure of subcenters associated with highway interchanges and airports in formerly rural areas, the so-called edge cities that came to our attention in the early 1990s.[1] The grid pattern of many older cities has given way to hierarchical street networks composed of sleepy residential crescents, collector through-streets, busier arterials, parkways, and highways, mostly without sidewalks. Older cities that were once dominated by precincts with a fine-grained mix of land uses—commercial, industrial, and residential areas located cheek by jowl—have evolved into metropolitan jigsaw puzzles of large-scale, single-use areas such as shopping malls, office parks, residential quarters, and entertainment complexes.

As we saw in chapter 4, Canadian cities show fewer signs of the stresses and strains typically associated with urban sprawl and car dependency. This suggests that if we look more closely at the urban form and transport systems of urban regions in the two countries, we might find some significant differences on the ground that would help explain Canada's better performance on sustainability and livability measures. To investigate this possibility, we start with an example of how the shape of city regions differ in the two countries; then, using data from urban regions across the two countries, we tease out the differences between the two countries in a more technical way.

Seattle and Vancouver

Seattle and Vancouver are two city-regions that invite comparison in that they have many attributes in common but they differ in crucial ways relevant to our inquiry. Only 150 miles apart, they are both coastal cities in the so-called Cascadia bioregion, with similar climate, flora, and fauna, surrounded by good-quality agricultural land, and cupped by the same majestic mountain range. They are both midsized urban regions: Vancouver with a population of 2.3 million and Seattle (including Tacoma) with 3.6 million people. The growth rate in both regions is very rapid at about 13 percent growth in Seattle in the first decade of the twenty-first century compared with 16 percent in Vancouver. At the heart of each region is a central city of just over 600,000 residents. Both central cities are laid out on a regular street grid; areas beyond the city boundaries have the familiar crescent and cul-de-sac street patterns that characterize suburban areas throughout both countries.

Despite these similarities, there are some key differences between the two regions in terms of urban form and transportation networks. As can be seen

TABLE 4-1.
Population densities (2010/2011) and transport system characteristics (2009) for Vancouver and Seattle regions

	Central Cities		Metropolitan Region					
Region	**Popul-ation**	**Density (people/ sq mile)**	**Popul-ation**	**Density (people/ sq mile)**	**Single-Family Homes as % of Total Housing**	**Public Transit (miles/ miles**	**Annual Vehicle Revenue Miles (miles/ person)**	**High-ways (lane miles per 10,000 popul-ation)**
Vancouver	603,502	13,590	2,313,328	120	33.8	5.4	40.5	1.8
Seattle	608,660	7,774	3,500,026	47	59.9	1.0	22.3	6.1

Sources: US Census Bureau, Statistics Canada Census; Green City Index, 2010; Margaret Ellis-Young and Craig Townsend at Concordia University, 2015.

from table 4-1, Vancouver's population density is much greater than Seattle's at both the city and regional scales. Although of similar population size, the City of Vancouver's spatial footprint is about half the size as Seattle's, so densities are almost twice as high.

These density differences illustrate the style of growth each area has seen over the last several decades. In Vancouver, the central city has pushed growth vertically, with residential high-rises throughout the downtown and a variety of housing types and commercial uses along arterial streets in the rest of the city. Much of the city is easily walkable, and virtually the whole city is dense enough to support good-quality transit. Seattle, on the other hand, has a downtown that is mostly office buildings surrounded by low-density residential neighborhoods, making downtown street life less vibrant than it would otherwise be, especially at night. Over the six decades from 1951 to 2011, the city of Vancouver steadily increased its population, ending the period with 75 percent more people than when it started. Meanwhile, Seattle added only 30 percent, mostly since 1990. Seattle's inner city has been mushrooming lately, but there are still only 80,000 people living within a mile and a half of the city's central point; in contrast, Vancouver has almost 130,000 residents within the same radius.

In the wider metropolitan regions, we find the same story. Although both regions have plenty of classic urban sprawl, there is much less of it in

Vancouver. Vancouver has tended to concentrate high-density development around its transit stations and has otherwise built many compact neighborhoods that can support frequent transit service. A variety of housing types can be found throughout the region, as reflected in the fact that only one-third of the housing stock being made up of single-family homes. Greater Seattle, in contrast, has grown outward, eating up farmland with a preponderance (60 percent) of single-family homes. On the whole, the Vancouver region is two and half times as dense as the Seattle region, with 120 versus 47 people per square mile.

A 2002 study by the Sightline Institute in Seattle (a nonprofit research group; formerly Northwest Environment Watch) compared growth patterns and density differences in the two regions.[2] It used the commonly accepted threshold of twelve people per acre to distinguish auto-dependent from transit-supportive neighborhoods; below that figure, the only feasible way to get around is by car, and above that figure, transit becomes cost-effective.[3] The authors reported that at the turn of the millennium, fully 62 percent of greater Vancouver's residents lived in transit-supportive neighborhoods. In contrast, only 25 percent of Seattle-area residents lived in such neighborhoods. Of greater Vancouver's residents, 11 percent lived in highly compact, pedestrian-oriented neighborhoods. Defined as forty residents or more per acre, these neighborhoods are places where up to one-third of households do not even own a car and where those who do tend to so use them relatively rarely. In contrast, only 3 percent of Seattle residents lived in such neighborhoods. Greater Seattle's development over the 1990s covered roughly twice as much land per new metropolitan resident as did greater Vancouver's. The density differences between the two regions can be clearly seen in figure 4-1.

The urban form differences between the two regions are complemented by differences in transportation systems. Greater Seattle has a network of major highways that promotes car dependence and development on the urban fringe, whereas greater Vancouver has a much more modest freeway system. The difference is especially stark when focusing on the central cities themselves. The City of Vancouver has no major highways, whereas the city of Seattle has two major north-south freeways going right through its center and several east-west connectors linking it to suburban municipalities. In total, the Vancouver region has only 1.8 lane-miles of freeways per 10,000 population; Seattle has 6.1.

(a)

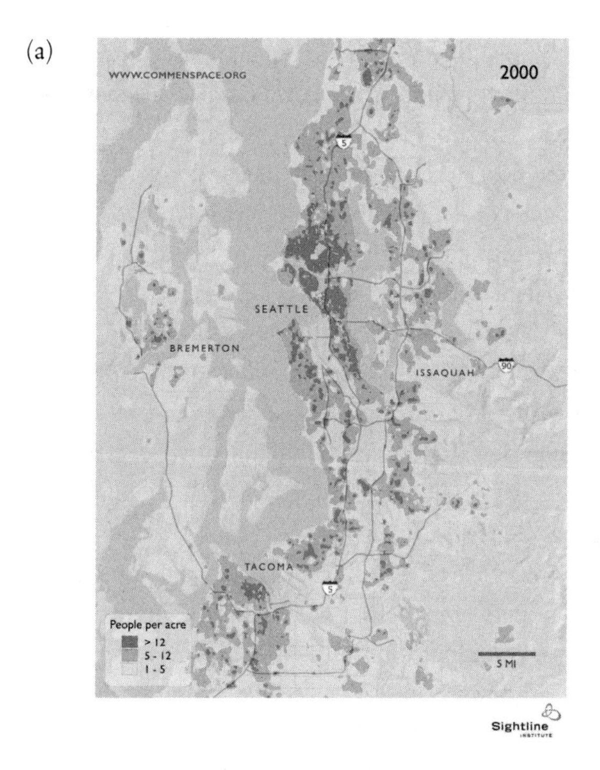

(b)

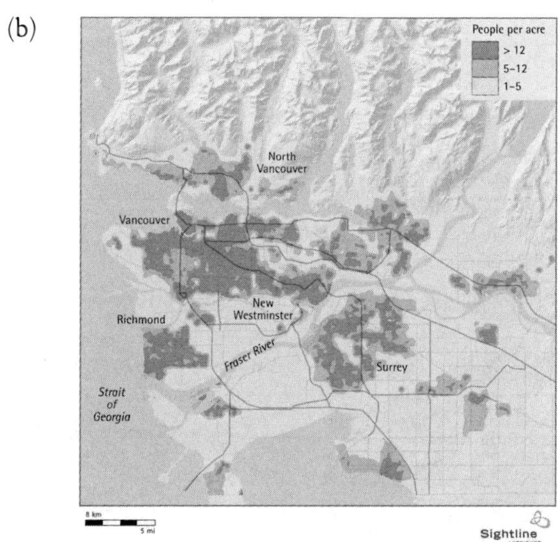

FIGURE 4-1. Density maps, (a) Seattle (2000) and (b) Vancouver (2001) metro regions

Source: Northwest Environment Watch and Smart Growth BC, 2002.

As for public transit, the Vancouver region boasts an extensive network of mostly elevated rapid transit routes—called SkyTrain—that links with a dense web of bus routes. The rapid transit network evolved over a thirty-year period and helped shape new urban growth or transform existing areas, with many stations at the heart of dense, mixed-use neighborhoods. Meanwhile, Seattle has only two recently built and fairly short rapid transit routes and an inadequate bus system with infrequent service outside of rush hours. Glancing back at table 4-1 we see the difference between the regions captured in the two transportation indicators: Vancouver's transit network is more than five times as dense as Seattle's and offers almost twice the level of service per capita. Clearly, the higher densities found in the Vancouver region have helped support a more extensive public transit system than that found in Seattle.

So how do differences in urban form and transportation systems affect sustainability outcomes? As table 4-2 reveals, Vancouver far exceeds Seattle on key sustainability indicators. If we look at the breakdown in the share of the various transportation modes in the two regions, we see that Vancouver has more than twice the transit mode, twice the biking share, and a 60 percent more walking share than the Seattle region. Greenhouse gas emissions are correspondingly lower in Vancouver, with only 7.1 metric tons per person emitted compared with 9.6 in the Seattle region. Particulate matter (and other air pol-

TABLE 4-2.
Sustainability indicators for Vancouver and Seattle regions

	Modal Share (%)				CO_2	Particulate Emissions (PM10)	Electricity Consumption (city only)	Water Consumption (city only)
Region	Car	Transit	Walking	Cycling	Metric Tons/ Person/ Year	Pounds/ Person/ Year	Gigajoules/ Person/Year	Gallons/ Person/ Day
Vancouver	70.8	19.7	6.3	1.8	7.1	7	32.5	121.9
Seattle	85.3	8.5	3.9	0.9	9.6	22	59.3	117.5
Difference (%)	–20	132	62	100	35	214	82	4

Sources: US Census Bureau, Statistics Canada Census, Green City Index 2010, and Metro Vancouver Integrated Air Quality and Greenhouse Gas Management Plan 2011.

lutants not shown in the table) in Vancouver is a fraction of what it is in Seattle. At the city level, electricity consumption is 82 percent higher in Seattle, with far more single-family homes to heat and cool, but per capita water consumption is about equal in the two cities.

Now we turn to the exploration of the various aspects of urban form and transportation systems on a wider basis for cities across the United States and Canada. We investigate a number of dimensions, including the following:

- Density: The number of people living on a given amount of land.
- Building form: The pattern of building types in the urban region.
- Land use mix: The degree to which different land uses (stores, parks, offices, etc.) are mixed together in a given area.
- Urban structure: The degree to which the urban region is centralized or decentralized.

Density

Urban densities are important from both environmental and livability points of view. Accommodating a given population on a smaller geographical footprint means less urbanized land, shorter travel trajectories, less energy usage for heating and cooling buildings, and lower infrastructure needs. Density can also add to the vitality of urban centers, contribute to the use of active transportation modes, and support investment in good-quality transit connections. Research shows that higher densities may work on their own or in combination with other factors and policy measures to produce these benefits. For example, the energy used for heating and cooling can be lowered by increasing community densities alone,[4] but lowering the amount of driving people do is best accomplished by raising densities in combination with other measures, such as creating new employment opportunities near where people live or restricting investment in new highways.[5]

Despite the importance of density to the sustainability of urban regions, getting an accurate measure of regional densities can be tricky, mostly because many rural or otherwise lightly inhabited places within the boundaries of a metropolitan region skew any density calculation that divides population by total land area. Different researchers overcome this handicap in different ways, which means that there are conflicting density measures for any given region. The general picture that emerges from density studies comparing US and Canadian urban regions transcends these difficulties, however, and that is the picture we paint here.

With the extreme compactness of Manhattan, the New York metropolitan area is by most measures the densest urban region in Northern America. Outside that one region—for which there is no counterpart in Canada—the density numbers point to a continuum of densities across Northern America, with urban regions like Los Angeles, San Francisco, and Chicago vying with Montreal, Toronto, and Montreal for the top spots. From this observation, some observers gather that US and Canadian urban regions are not very different in terms of the overall densities. A look below the surface, however, reveals important distinctions between the two countries.

Because density varies with city size, larger urban regions are expected to have higher densities; the extreme is, as we've just pointed out, New York. When Canadian and US urban regions with similar populations are placed side by side, Canadian cities are clearly denser than their US counterparts. For example, the University of Toronto's John Miron found that for urban regions of middling sizes (100,000 to 1 million or 1 million to 4 million), Canadian metros are about twice the density of their US peers. For cities with more than 4 million persons, Canada's lone contender—Toronto—is 50 percent denser than the average of six US metros in that size category (e.g., Los Angeles, Chicago, San Francisco), not counting New York City.[6] Another way of expressing this difference is to say that modestly sized Canadian urban regions—places most Americans have never heard of—are denser than much larger US regions. For example, the Hamilton region, west of Toronto in Ontario, had a population of only 624,000 as of 2001 but was denser than Philadelphia (5.9 million), Boston (4.1 million), and Washington, DC (3.9 million). Other studies have confirmed this pattern.

Zachary Taylor at the University of Toronto has used a simple index to compare the overall densities of US and Canadian metropolitan areas, regardless of size. His "sprawl index" goes from 0 to 100, with 0 meaning that the whole metro population lives in high-density census tracts and 100 meaning that the whole population lives in low- density tracts. The results for census years 2000 in the

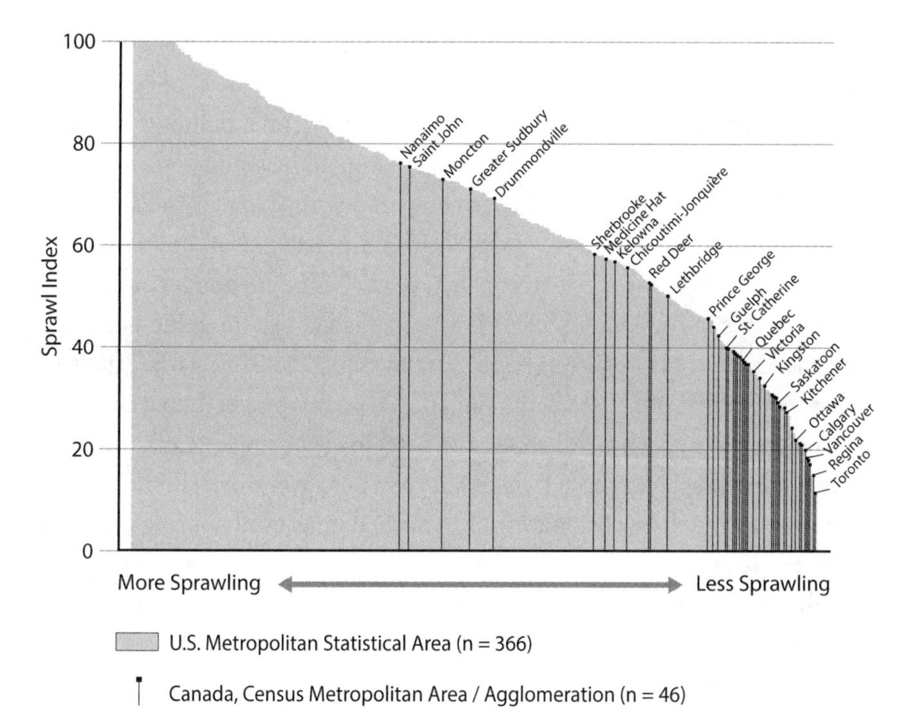

FIGURE 4-2. Sprawl index, all metro areas in United States (2000) and Canada (2001)
Source: Taylor 2015.
Note: For readability, only selected Canadian cities are labeled.[7]

United States and 2001 in Canada are shown in figure 4-2, with each vertical line representing a different metropolitan area. Gray lines are US metro areas, and black lines are Canadian metro areas. Most Canadian metropolitan areas, and especially the large ones, have low scores, meaning that a greater proportion of metropolitan residents in that country live in high-density census tracts.

Building Form

If it is true that Canadian cities are generally denser than their US counterparts, you would expect to find this fact reflected in the mix of building types found there. For instance, there might be fewer single-family detached homes than in the United States and more tall buildings. Indeed, that is exactly the case.

Single-family detached homes—the type of residential building form that takes up the most land and uses the most energy to heat and cool—are relatively abundant in the United States compared with Canada. In 2010–2011, single-family detached units accounted for 63 percent of all housing in the United States but only 55 percent in Canada.[8] The preponderance of detached homes in the United States is an established pattern going back to at least the 1950s. This trend can be seen by tracking "housing starts," statistics about the number of houses being built of different types, collected each year in both countries. Figure 4-3 shows the percentage of single-family detached houses started each year from 1959 to 2014 as a percentage of all housing starts. It shows that the share of single-family detached units of all housing starts was significantly greater in the United States than in Canada for almost every year over that 55-year period. In recent years, the spread between the two countries

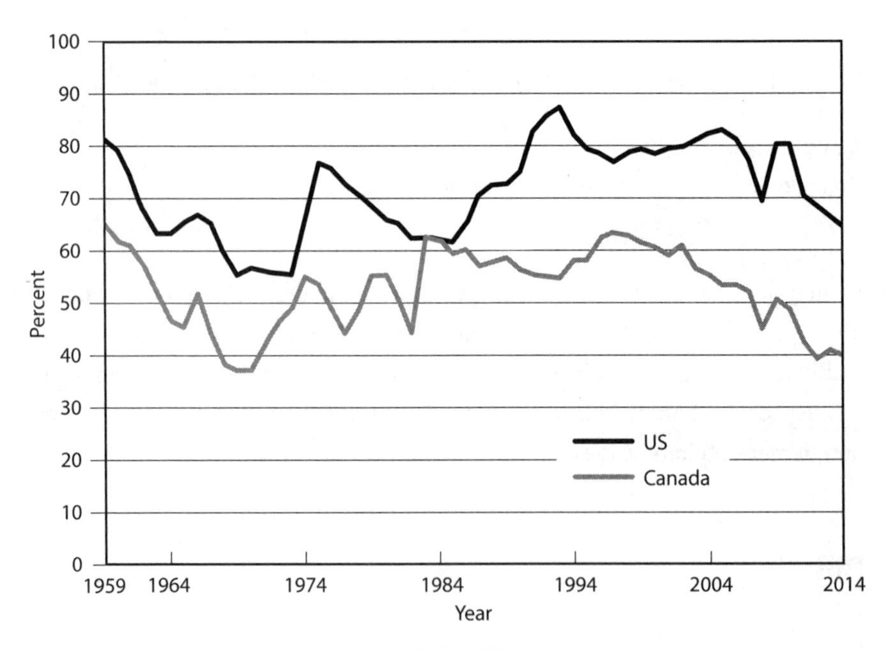

FIGURE 4-3. Single-family detached housing starts as percentage of total housing starts, United States and Canada, 1959–2014
Sources: US Census Bureau; Statistics Canada Census.

appears to have widened somewhat, with a major decline in Canada setting in since the mid-1990s compared with a smaller decline in the United States. In the United States, 71.2 percent of the housing units started from 1959 to 2014 were single-family detached units; in Canada, the figure was just over half (52 percent).

At the other end of the density spectrum in terms of building form are high-rise buildings. Anecdotal observation of Canadian cities suggests that they tend to have more high-rise structures than their US counterparts and that these buildings tend to be spread throughout the urban area rather than concentrated in the downtown. In Canada, it is not unusual to see concentrations of high-rise buildings in suburban municipalities, especially where good-quality transit is available. Mississauga, for example, grew as a bedroom community to Toronto but started to intensify in the 1990s when it ran out of greenfield opportunities. Although it still has plenty of single-family housing, the municipality now boasts dozens of high-rise buildings in its burgeoning city center (figure 4-4).

FIGURE 4-4. Absolute Condos, Mississauga, Ontario
Source: Terry Ozon, flickr.

The website Skyscrapers.com tracks skyscrapers throughout the world, logging the size, construction date, use, and location of buildings that are more than twelve stories tall. Data from the site show that Canadian central cities have considerably more tall buildings than their US counterparts. Aside from New York City, Toronto has the most high-rises of any city in either country. Toronto and Chicago have about the same populations, but the former has 2,039 tall buildings, whereas the latter has only 1,154, about half as many. Montreal has 630 high-rises, whereas Philadelphia, with a comparable city population and a much larger metropolitan population than Montreal's, has only 367. In Vancouver, there are 674 tall buildings, whereas Seattle, which is about the same size, has only 234.

Of the top fifty cities in either country ranked according to the number of high-rises, thirteen are in Canada, far exceeding its population weight relative to the United States, and thirty-seven are in the United States. The average city sizes are about the same between the two sets of cities, but the number of high-rises in the Canadian set of cities is almost double that in the United States: 409 versus 227 (table 4-3).

By comparing maps on Skyscraper.com that show the location of high-rises in each city, the user can get a good snapshot of the differences between the two countries. Figure 4-5 highlights the typical distribution of high-rises in Canadian versus US cities, showing the locations of high-rise buildings in Montreal and in Philadelphia. Not only does Philadelphia have fewer high-rise buildings than Montreal, but they are concentrated in the city center and along a limited number of spines. High-rises in Montreal are more plentiful than in Philadelphia and are spread throughout the urban area.

TABLE 4-3.
Average number of high-rises in top fifty central cities

Country	Number of Cities in Top Fifty Cities Ranked by Number of High-Rises	Average Number of High-Rises per City	Average City Population
Canada	13	409	826,191
United States	37	227	818,714

Source: Skyscrapers.com.

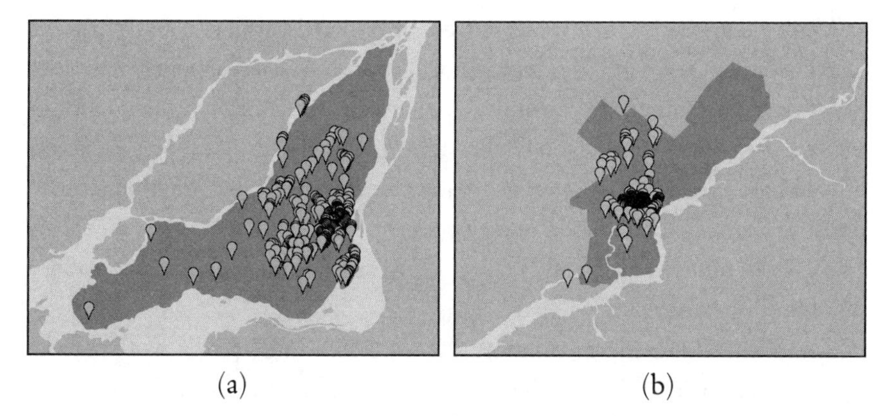

FIGURE 4-5. (a) Montreal versus (b) Philadelphia high-rise building locations
Source: Skyscrapers.com.

Land Use Mix

Mixing land uses—whether it be within a building, on a street, in a neighborhood, or on an even wider scale—affects travel behavior because it increases the availability of destinations within a given area. Increased mix reduces travel distances and allows more transit, walking, and cycling trips as opposed to trips in private vehicles. Planners are increasingly interested in creating "complete communities," in both urban and suburban areas, that combine daily destinations, such as stores, schools, and parks.

Land use mix can be measured at a variety of scales, but it is most salient at the neighborhood level, where increasing the mix of various land uses can significantly boost walking and biking. The "walkability" of urban districts is increasingly associated with both a high quality of life and environmental sustainability. People living in walkable, mixed-use neighborhoods are more likely to know their neighbors, participate politically, trust others, and be involved socially.[9] A walkable neighborhood tends to have lower levels of obesity and overweight residents as a result of higher levels of physical activity than do nonwalkable neighborhoods. Walkability facilitates walking, not only to final destinations but also to transit.[10] Thus, walkability is a complement to transit access, mixed land uses, and higher densities, which, taken together, form the core of urban planning efforts to reduce auto dependence.[11]

Walk Score is a measure of land use mix that is available for many US and Canadian cities.[12] The Walk Score algorithm awards points based on the distance to the closest amenity in each land use category, such as businesses, parks, theaters, schools, and other common destinations. Points are also awarded based on pedestrian-friendly features such as higher population density, shorter block length, and greater intersection density. In this way, a Walk Score from 0 to 100 can be assigned to any address and, by extension, a neighborhood or even a whole city.

The Walk Score website has scores for Canadian and US cities with populations of more than 200,000. The scores put Canadian and US cities on a continuum, with New York City, Jersey City, San Francisco, and Boston scoring higher than Vancouver, the highest scoring Canadian city. A closer look reveals some interesting patterns, however. The lowest score among the Canadian municipalities was 43.8 (Gatineau, Quebec), whereas the lowest score among the US municipalities was 18 (Fayetteville, North Carolina). Moreover, several of the municipalities in Canada with respectable Walk Scores are suburban jurisdictions, far from the centers of their urban regions. For example, Mississauga—which is almost 20 miles from downtown Toronto, scored 58.6, just below Milwaukee. Surrey is even farther from the center of the Vancouver region, but it achieved a Walk Score of 51.2, just below Detroit. Longueuil, which is not even on the island of Montreal, scored 56.4, just above New Orleans.

The Walk Score averages for cities with populations of more than 200,000 are shown in table 4-4. For the US cities, the average Walk Score is 45, whereas for the Canadian group, the average is 54, or 20 percent higher. According to the Walk Score website, a score above 50 corresponds to an urban form that is "somewhat walkable," whereas a score below 50 is "car-dependent." The

TABLE 4-4.
Average walk score in cities with more than 200,000 population, Canada and the United States, 2015

Country	Number of Cities	Average Walk Score	Average Population
Canada	22	54	627,101
United States	108	45	569,526

Source: www.walkscore.com.

significant difference in the two countries suggests that Canadian cities have on average a greater degree of land use mixing at the neighborhood scale, which contributes to a more walkable urban form.

Urban Structure

The structure of a city and its region has a significant effect on its sustainability performance. Cities with concentrations of higher density and multifunctional land uses tend to be more efficient in the provision of public services. For example, public transport can be more efficiently organized when people and jobs are concentrated in centers of a certain size, which ensures the achievement of economies of scale. There are two ways to compare Canadian and US cities in terms of urban structure: via density gradients and by looking at central city populations.

Density Gradients

The density gradient is a useful snapshot of how the population of a city is distributed as one moves from the center to the periphery. A steep gradient means that the city has a dense center and a compact periphery. A low gradient implies that the city is spread out rather homogenously. If measured at different times, density gradients can reveal whether a region is decentralizing and, if, so, at what rate.[13] The density gradient is usually measured along with central-area densities because, depending on how high the "starting" density is in the city center, cities with similar density gradients could have substantially different densities.

Pierre Filion and his colleagues at the University of Waterloo used density gradients to compare the centrality of urban regions on the two sides of the border. They tracked density patterns of the three largest Canadian metropolitan regions (Montreal, Vancouver, and Toronto) and compared them with those of a sample of twelve US urban areas with comparable populations.[14] Using 1990–1991 data, they grouped the study cities into three categories based on their density gradients: centralized metropolitan regions; weakly centralized

TABLE 4-5.
Zonal density (people per square mile), Canada (1991) and the United States (1990)

	Entire Built-Up Area	Core Areas	Inner Cities	Inner Suburbs	Outer Suburbs
Canadian average	6,348	16,687	15,833	7,806	4,219
US average	4,188	8,138	7,014	4,069	3,305

Source: Filion et al. 2004.

metropolitan areas; and dispersed, low-density regions. All three Canadian cities but only one-third of the US cities (Boston; Philadelphia; Minneapolis; and Washington, DC) fit into the centralized metropolitan region category, with high central-city densities and steep declines in density levels moving outward. The remaining eight US cities all had relatively flat density gradients. Montreal had the steepest gradient of the fifteen metros in the sample, followed by Toronto and Vancouver. Although not entirely in a category of their own, Canadian cities, it would seem, tend to be more centralized than their US counterparts.

Filion and his colleagues also compared densities in different zones of the built-up area for each urban region by separating out census tracts in the core area and three other zones defined by the historical time of development: the inner city defined as mostly pre-1940, inner suburb defined as mostly 1940–1970, and outer suburb defined as post-1970.[15] The results (summarized in table 4-5) showed that the Canadian cities were on average far denser than the US cities in every "time zone," although the differences in the outer suburbs were not as dramatic as they were in the other zones. This finding would suggest that residential building trends may be converging somewhat in newly built suburbs.

Central-City Populations

A good indicator of the relatively healthy state of central cities can be found in population trends. Struggling central cities are characterized by high crime rates, concentrated poverty, physical decay, and poor fiscal conditions. They tend to bleed population as better-off residents flee deteriorating conditions, contributing to the "flight-from-blight" cycle in the city and sprawl in the suburbs.[16] Healthy central cities are able to maintain services and continue to offer a good quality of life, preserving existing residents and attracting new migrants. A strong central city can also serve as the hub of an efficient regional transportation system.

Population trends in Canadian central cities can be tricky to track over time because these cities often annex or amalgamate with surrounding suburban communities and therefore change their geographic definition. Building on the work of Michael Lewyn at the Touro Law Center, we addressed this problem for the largest Canadian cities by tracking population changes in those census tracts that defined central cities as of 1971 and ignoring the population in geographic additions after that year.[17] The results are presented in table 4-6. The table shows that seven of the ten Canadian cities gained population from 1971 to 2011, with gains ranging from 2.4 percent in Toronto to 168 percent in Calgary, whereas losses in the other three cities ranged from 10.3 (Quebec) to 16.5 percent (Montreal). The average gain for the 1971 to 2011 period was 24.8 percent. If we look at the five cities for which we have data for the 1951–2011

TABLE 4-6.
Population trends, largest Canadian cities, 1951–2001

	Population (thousands)				Change (%)				
City	1951	1971	2001	2011	1951–2011	1971–2011	1951–1971	1971–2001	2001–2011
Calgary		403	879	1081		168.0		117.9	23.0
Edmonton		438	633	680		55.2		44.5	7.4
Hamilton		309	331	330		6.9		7.1	–0.2
Montreal		1214	991	1014		–16.5		–18.4	2.3
Ottawa	202	302	337	338	67.2	11.7	67.2	11.5	0.2
Quebec	164	186	159	167	1.8	–10.3	1.8	–14.5	5.0
Toronto	676	713	676	730	8.0	2.4	8.0	–5.1	7.9
Vancouver	345	426	546	604	75.0	41.6	75.0	28.0	10.6
Windsor		203	208	210		3.4		2.5	0.9
Winnipeg	236	246	207	211	–10.4	–4.2	–10.4	–6.0	2.1
Average					28.3	24.8	28.3	15.4	5.9

Source: Courtesy of Michael Lewyn; Statistics Canada.

TABLE 4-7.
Population trends, largest US cities, 1950–2000

	Population (thousands)				Change (%)				
City	1950	1970	2000	2010	1950-2010	1970–2010	1950–1970	1970–2000	2000–2010
Baltimore	949	906	651	621	–34.6	–31.5	–4.5	–28.1	–4.6
Boston	801	641	589	618	–22.8	–3.6	–20.0	–8.1	4.9
Chicago	3,620	3,367	2,896	2,695	–25.6	–20.0	–7.0	–14.0	–6.9
Cleveland	914	751	478	397	–56.6	–47.1	–17.8	–36.4	–16.9
Detroit	1,849	1,511	951	713	–61.4	–52.8	–18.3	–37.1	–25
Los Angeles	1,970	2,816	3,694	3,793	92.5	34.7	42.9	31.2	2.7
New York	7,891	7,895	8,008	8,175	3.6	3.5	0.1	1.4	2.1
Philadelphia	2,071	1,949	1,517	1,526	–26.3	–21.7	–5.9	–22.2	0.6
St. Louis	856	622	348	319	–62.7	–48.7	–27.3	–44.1	–8.3
Washington, DC	802	757	572	602	–24.9	–20.5	–5.6	–24.4	5.2
Average					–21.9	–20.8	–6.3	–18.2	–4.6

Source: Courtesy of Michael Lewyn; US Census Bureau.

period, only Winnipeg lost population over that extended time period. Ottawa, Vancouver, and Toronto gained population, and Quebec stayed about level. The average gain for that longer period was 28.3 percent.

For comparison purposes, the data in table 4-7 show population trends from 1950 to 2010 for what were the largest US central cities in 1950. Clearly, US central cities have not fared as well as their Canadian counterparts. Eight of the ten largest cities lost population from 1950 to 2000, and one (New York) experienced a population gain of only 3.6 percent. The only city that grew substantially, Los Angeles, had a vast amount of undeveloped land within its 1950 city limits and was thus able to sprawl without losing overall population. Losses in US cities from 1970 to 2010 were in the range of 3.6 percent (Boston) to 52.8 percent (Detroit), with an average population change of –20.8 percent across all ten cities.

The figures in tables 4-6 and 4-7 suggest—consistent with conventional wisdom—that Canadian central cities are in a healthier state than their US counterparts. Moreover, there is evidence that the more modest depopulations that have been experienced in Canada are of a less serious type than seen in US cities. Most of the distressed central cities in the United States have not only shed population but also lost households, which can be especially devastating because it implies abandonment—even demolition—of deteriorated buildings. For the period from 1981 to 2001, none of the Canadian cases involved losses in the number of households, which suggests that central cities that lost population did so due to a drop in the number of people per household (household size), a universal trend that need not pose a challenge to neighborhood stability.[18]

Transportation Infrastructure and Service

Transportation infrastructure affects sustainability and livability both directly and indirectly. The direct effects arise simply because different types of transportation infrastructure encourage different types of travel behavior, which in turn have different environmental and social effects. Building multilane roads encourages people to drive (fast) rather than bike or walk and thus raises the environmental footprint of the community and reduces its livability. The indirect effects of transportation infrastructure arise because such infrastructure can encourage changes in land use patterns by influencing the relative accessibility of land in different locations, which in turn promote shifts in travel activities.

There is an increasing awareness of how the urban form and transportation dimensions of the urban system can be mutually reinforcing so as to push the city toward more sprawl or toward smarter growth. To take one of the most discussed examples of circular causation, let us consider a new arterial road on the urban fringe. The road reduces travel times by car and opens up a new area for development designed around car use, undermining transit ridership and the viability of alternatives to the car, causing congestion, and leading to demands for more roads, thus introducing another turn of the cycle. In general, automobile-oriented transport planning tends to cause more dispersed, automobile-oriented development (sprawl). Walking and transit improvements tend to have opposite effects, encouraging more compact, mixed, and multimodal development.[19]

Obviously, an automobile-oriented city will have more highways, roads, and parking lots, and a non-automobile-oriented city will have more transit stations, rail corridors, bus lanes, bike lanes, and sidewalks. The difference in energy consumption and greenhouse gas emission between these two types of cities can be substantial. Also of note are the effect of impervious road surfaces and parking lots on surface and groundwater flows and the fragmentation of natural areas by highways and major roads. Investments in different types of transportation infrastructure can also affect social equity. For example, new roads and highways tend to favor those who own and use cars the most, whereas the benefits of investments in transit, walking, and biking are more evenly spread out.[20]

Highways

Highways, a word often used interchangeably with freeways or expressways, are roadways designed to move large volumes of automobiles at high speeds. Evidence shows that highway investments in urban areas can have a significant effect on the shape of the region.[21] By reducing the cost and time of travel, highway construction has made it possible for workers to live farther from central cities, a trend that is often followed by a decentralization of major employers. By creating high-value intersections throughout the suburban periphery, highways attracted shopping malls and power centers, undermining the strength of central cities as retail centers. When running through urban areas, highways have disrupted the previously tight-knit fabric of cities, undermined the viability of otherwise healthy neighborhoods, created noise and pollution, and made cities less desirable places to live. Freeways have also shifted travel demand from transit to private vehicles as revealed by cities with expansive highway systems usually having lower levels of transit provision and ridership.[22] The overall judgment is that highways have contributed heavily to the decline of city centers, transferring much of the city's vital energy to the urban fringe and promoting car dependence.[23]

On the symbolic level, highways are an expression of a city's priorities in terms of the provision of transport infrastructure. The sprawling, car-dependent city of San Jose, California, has an interlacing highway network and an inadequate public transit system. In contrast, Vancouver consciously chose to cancel major freeway plans of the 1960s and remains one of the only major cities in Northern America with no highways within the core area. It is widely associated with walkable neighborhoods; is ranked as one of the most livable cities in the world; and retains a highly functional, diversified passenger transport system.

Although San Jose and Vancouver may present extremes of the spectrum, they do in fact reflect differences between the two countries in terms of urban freeways, a difference that has persisted for decades. In 1985, Barry Edmonston, Michael Goldberg, and John Mercer computed the number of expressway lane-miles per capita and found that there were four times as many lane-miles of urban expressway for each metropolitan resident in the United States as there were in Canada: 1.3 lane-miles per thousand people in the United States versus 0.3 in Canada.[24]

Thirty years later, a study by Craig Townsend and Margaret Ellis-Young at Concordia University in Montreal showed that Canadian cities continue to have fewer expressways compared with their US counterparts. The authors collected data on highway lane-miles for all fifty-seven metropolitan areas in Northern America with more than one million inhabitants. After excluding New York, Los Angeles, and Chicago—three regions that are far larger than any Canadian counterpart—they ranked urban regions according their lane-miles per capita. Focusing on the city centers, they found that five of the six Canadian cities fell into the lowest 20 percent of cities ranked this way. Edmonton, Calgary, and Vancouver did not have any freeways at all within the central area, a feat unmatched by any of the US cities. At the metropolitan scale, the Canadian contenders were again distinguished by their low level of freeway lane-miles compared with the US regions in that all six were ranked in the lowest quintile.

Roads and Parking

The size of the road network and amount of parking also affect travel behavior and urban form. An extensive network of roads and high levels of free, easily accessed, vehicle parking encourage greater levels of car travel. Particularly in the central business districts, parking provision is an important factor in

TABLE 4-8.
Road lengths and parking spaces in a sample of US and Canadian metro areas, 1995 and 2005

	US Cities			Canadian Cities		
Variable	**1995**	**2005**	**Change (%)**	**1996**	**2006**	**Change (%)**
Length of road (feet per person)	21.3	19.7	–7.7	17.4	17.7	1.9
Parking spaces per 1,000 jobs in central business district	555.0	487.0	–12.3	390.0	319.0	–18.2

Source: Newman and Kenworthy 2015.

determining whether people commute by car or transit. Research shows that measures to limit parking are among the most effective methods to move mode choice away from cars.[25]

Peter Newman and Jeff Kenworthy have tracked transportation indicators for thirty-three cities around the world, including ten in the United States and five in Canada.[26] The data are available for both 1995 and 2005, providing some insight into trends over time in the two countries. Among the indicators included in their database are two of relevance here, shown in table 4-8: Canadian cities in the sample have fewer feet of roads per person, and they have significantly fewer parking spaces in their central business districts per 1,000 jobs.[27]

Transit Infrastructure and Service

The level of transit ridership in a city is determined by a number of factors both internal and external to the city's transit systems. Among the external factors, urban form is a major determinant of transit patronage. Higher densities make better quality transit economically feasible, whereas compact, mixed-use development around transit stations contributes to transit ridership. As we have already seen in this chapter, densities and land use mix tend to be higher in Canadian cities, factors that no doubt contribute to the higher transit ridership in that country.[28]

TABLE 4-9.
Road lengths and parking spaces in a sample of US and Canadian metro areas, 1995 and 2005

	US Cities			Canadian Cities		
Variable	**1995**	**2005**	**Change (%)**	**1995**	**2005**	**Change (%)**
Public transport Infrastructure factors						
Total length of reserved public transport route per person (feet/1,000 persons)	159.8	235.2	47.2	184.7	219.2	18.7
Public transport service						
Total public transport seat miles of service per person	969.3	1,164.4	20.1	1,422.9	1,471.4	3.4
Total rail seat miles per person	469.8	625.1	33.1	420.0	522.6	24.4
Total bus seat miles per person	499.6	531.3	6.3	998.5	945.7	–5.3

Source: Newman and Kenworthy 2015.

Factors internal to transit systems include the reliability and frequency of service, comfort, spacing of transit routes, and the speed of transit services. Newman and Kenworthy's data set includes several indicators that can help us compare the two countries along these lines. Reserved public transport routes, such as bus lanes and rail systems operating on their own dedicated rights-of-way, are important for ensuring the reliability and speed of transit services. The figures in table 4-9 show that in 1995, Canada had a greater length of such routes per 1,000 people, but lost this edge to US cities in 2005.

Despite having lower levels of dedicated transit routes, the other measures reported by Newman and Kenworthy show that the Canadian cities in the sample have, on average, far superior levels of service compared with US cities.[29] The number of miles traveled by all transit vehicles per capita is significantly higher in Canada than in the United States. Seat-miles, which is the vehicle miles multiplied by the average numbers of seats in transit vehicles, are also much higher in Canada than in the United States. When broken down by rail versus bus, it is clear that Canada's preeminence in this regard is limited to bus systems. In fact, the US cities appear to have more rail vehicle seats per capita than do Canadian cities, reflecting the heavy investment in light rail transit and commuter rail in the United States from 1995 to 2005.

Although Canada is ahead of the United States on most measures of transit service, table 4-9 also shows that the gap between the two countries is closing, with US cities in the sample making more impressive gains in their service levels than did the Canadian cities. It is important to note, however, that higher service levels do not necessarily translate into higher ridership if external conditions do not support travel behavior change. In particular, if densities are too low or if car ownership rates are high and cost of use is low, increasing service levels may have little effect on transit patronage. We discuss this subject at greater length in chapter 7.

Bike Infrastructure

By European standards, neither Canadian nor American cities are particularly bike-friendly; the car is king almost everywhere in Northern America. As we noted in chapter 3, however, Canadian cities have, on average, about three times the rate of cycling to work as do US cities. What explains this cross-national difference? Empirical research has shown that biking levels are influenced by the kind of urban form and transportation characteristics discussed above. The denser urban fabrics found in Canadian cities tend to make trips shorter, which favors the use of active transportation modes such as walking and biking. The greater mixing of different land uses also makes destinations more accessible in Canadian cities and contributes to higher cycling rates. The more restrictive car parking policies in Canadian cities are not intended to encourage cycling, but they may have that effect.[30]

Another factor that has been shown to boost biking is the provision of cycling facilities, such as bike lanes and bike parking. Measurements collected by Meghan Winters at Simon Fraser University allow us to compare bike facilities in ten Canadian cities with populations ranging from 56,000 (Fredericton, New Brunswick) to 2.6 million (Toronto) with those in seventeen US cities, the smallest of which had a population of 97,000 (Boulder, Colorado) and the largest of which had 8.2 million (New York) people. The data in table 4-10 are broken into two categories of bike facilities: bike lanes that are physically separated from traffic (e.g., with a concrete median) and bike lanes that are merely painted on the roadway. The figures show that the sample of

TABLE 4-10.
Cycling infrastructure for selected Canadian and US cities, 2015

Type of Cycling Infrastructure	Miles per 100,000 People, Canadian Cities	Miles per 100,000 People, US Cities	Difference (%)
Separated lanes	20.1	6.6	206
Nonseparated lanes	16.5	15.5	7
Total	36.6	22.0	66

Source: Courtesy of Meghan Winters, Simon Fraser University.

Canadian cities has about the same length of painted bike lanes as US cities, but three times more separated tracks.

To illustrate the differences between the two countries, we can compare bike infrastructure in Montreal and Philadelphia, two cities of about the same population size (around 1.6 million). The two cities have about the same length of lanes painted on city streets: 191 in Montreal and 212 miles in Philadelphia. The major difference between the two cities is in the length of separated paths: Montreal has 240 miles compared with only 70 in Philadelphia. Another difference between the two cities is that Montreal has separated bike lanes both on city streets and in off-street locations (such as through parks), whereas in Philadelphia, all the separated lanes are off-street. Bike lanes on city streets are more likely to boost commuting by bike because they are designed and located for utilitarian travel as opposed to recreational purposes.

The difference between separated and painted cycling facilities is important because research has shown that separated paths are safer than painted lanes and bikeways.[31] Moreover, most cyclists feel safer on separated facilities, and because the perception of safety is a major determinant of the decision to bike, facilities separated from vehicle traffic have a higher potential to generate new riders.[32] Finally, because they are many times more expensive per foot to build than nonseparated paths, separated paths may reflect a greater political commitment to supporting cycling.[33]

The different levels of separated bike facilities in the two countries not only help explain the higher bike ridership rates, but also help account for differences in the level of bike accidents. The United States has an extremely high rate of cycling fatalities—9.24 per 100 million miles cycled—compared with other industrialized countries. In contrast, Canada's fatality rate of 3.85 cycling fatalities per 100 million miles cycled is similar to that of European countries, such as Germany and France.[34]

The relationship between cycling safety and ridership is a self-reinforcing circle. Safer cycling encourages more people to cycle, and as more people cycle, there are more cycling facilities and more awareness among motorists of cyclists, making cycling safer.[35] Thus, Canada's relatively high bike mode share, greater number of bike lanes, and low cycling fatality rate may be functionally interrelated.

The availability of bike parking is another key issue that affects people's willingness to bike and shows that Canadian cities seem to be consistently ahead of their US counterparts. Toronto has about 20,000 post-and-ring dual parking racks on sidewalks and additional standard multibike racks at most subway and commuter train stations. Montreal has 12,000 spots on specially designed car parking meters and another 13,000 in racks on city streets and outside transit stations. Ottawa has more than 15,000 bike racks in public spaces and outside government offices. In contrast, most large US cities provide less bike parking than even medium-sized cities in Canada. In the United States, Chicago tops the list of bike parking with 14,500 bike parking spots on sidewalks, with New York as the runner up with about 12,000 sidewalk racks.

Conclusion

This chapter has explored differences between US and Canadian cities and metros on a physical level, including city morphology and transportation systems. The evidence presented here supports the argument that there are significant differences between the two countries. In terms of urban form, Canadian urban regions are denser, are less dispersed, have more stable central cities, have a great mix of land uses on a neighborhood scale, have fewer land-hungry detached dwellings, and have more high-rise buildings spread throughout the city than do their US counterparts. Moreover, it appears that at least some of these characteristics have differentiated cities in the two countries for many decades. From the point of view of transportation systems, Canadian cities have fewer highways, fewer roads, fewer parking spaces in the central area, better transit services, and more secure biking paths than cities in the United States. For most of these factors, we have shown that these differences persist even after taking into account variations in city size between the two countries.

These long-lasting contrasts help explain the sustainability and livability differences between the two countries that we noted in chapter 3. With more centralized, compact cities and fewer highways, it is easier to support better transit services. With such conditions, it is not surprising that Canadian urbanites own fewer cars, drive them less, and tend to use transit more than their US cousins. With less private and more shared transport activity, it follows that Canadian cities have higher grades on sustainability measures such as greenhouse gas emissions and energy use. Better biking infrastructure has undoubtedly contributed to a greater interest in two-wheeled transport by increasing both actual and perceived safety. Lower levels of motorized transport, less disruption to the urban fabric from introducing highways into the central city, and greater access to transit services have probably contributed to the relative stability of central-city populations. Less motorized transport and more walking and biking probably have an effect on public health conditions, in particular on the lower incidence of overweight and obese people in Canadian cities compared with their US counterparts.

Needless to say, it is very difficult to untangle cause and effect in many of these associations. Take, for example, the relative stability of populations in Canadian central cities compared with cities in the United States, where populations have often fallen, in some cases dramatically, as their metro areas grew. Stable central cities and the distribution of higher-density building forms on a wider basis in Canadian cities may be a result of or a contributor to better transit services. There may be fewer freeways in Canadian cities because there is less demand given the viability of other travel modes such as public transit, or fewer freeways may be causing more congestion and forcing people onto public transit. Untangling cause and effect in complex urban systems is an immensely complicated issue, and we make no pretense of being able to prove causal relationships. What we can say, however, is that many of the differences between Canadian and US cities are mutually reinforcing and seem to form a gestalt of features that have put Canadian cities on a more sustainable, livable path over several decades.

Our main interest is not with the intricate web of cause and effect among the variables of urban form, transportation systems, livability, and sustainability, but in trying to understand the larger factors that may have put Canadian cities on a more sustainable path in the years following World War II. In the search for explanations, some observers may be tempted to claim that the key differences reflect impersonal forces such as the small difference in average incomes between Canadians and their US brethren. Lower affluence might account for a range of factors such as lower levels of single-family home production, smaller building lots, lower levels of car ownership, and greater interest in transit, walking, and biking in Canada relative to the United States. Our position is that lower affluence in Canada should be acknowledged as a background factor to be kept in mind as we proceed, but it is unlikely to satisfy our quest to understand the significant and enduring differences between the two countries that we have observed.[36]

Rather than focusing on economic or technological factors, most analysts who have explored the "same but different" relationship between the United States and Canada have had their attention drawn to governance structures and public policies to explain the observed disparities. For example, in their discussion about metropolitan density differences between the two countries, Filion and his colleagues noted the interacting nature of strong central cities, better transit systems linking city centers to higher-density suburbs, and the relative paucity of highways in Canada versus the United States. They attribute these conditions to stronger regional governments in Canada with enough authority to influence development patterns, better coordination between land use and transportation planning, planning standards that support higher densities, and higher levels of immigration to prevent the demographic decline seen in US central cities. We will explore these and other issues related to the governance and policy differences between the two countries in chapters 5 through 7.

Chapter 5

Organizing Government: Powers, Boundaries, and Governance Systems

As we saw in earlier chapters, the United States and Canada share many similarities, including their patterns of urbanization over time, their federal form of government, their prosperity, cultural expression, and advanced economies. We also saw, however, that Canadian cities are more compact; have a greater mix of daily destinations within walking distance; and have more stable and healthier central cities, better transit systems, fewer highways, fewer roads, and more sophisticated cycling infrastructure than US cities. So why, if the two countries are so similar in so many ways, are their urban forms so different?

In this chapter, we begin to answer this question by looking at how the two countries differ in terms of the governance of urban regions. Our purpose is to show how government structures and the allocation of responsibility for matters related to the planning and management of urban regions affects how cities grow and how people get around in them. Our investigation covers such territory as the organization of local government (whether consolidated or fragmented across the urban region), the amount of local autonomy exercised by municipalities (whether they make planning decisions more or less independently or within a framework laid down by states or provinces), and the way urban regions are governed (whether they rely on loose collaboration or on more formal government structures). Our primary focus is on how these issues relate to land use planning. In the next chapter, we explore the implications for transportation planning in greater detail.

Municipal Fragmentation

As cities grow, the municipal organization of the area becomes ever more complex: to provide services to the growing population and employment base, previously independent towns may become part of a continuously urban region or unincorporated rural areas may be reorganized into new municipal units. Although the city-region may function as an integrated entity from a social and labor market point of view, the multiplication of local jurisdictions making up metro regions as they grow means that governance is increasingly fragmented among a number of independent local authorities. As we will see, the degree of fragmentation can have serious implications for how the urban area grows, affecting both its spatial and social structure.

Canadian metropolitan regions tend to have far fewer municipal governments than their US counterparts, a pattern that goes back to at least the 1970s, when Michael Goldberg and John Mercer calculated government densities to compare US and Canadian metropolitan areas. The metric reflects the number of municipalities in a metropolitan area per thousand population resident in municipally governed areas. The Canadian mean score on this index was substantially lower than that for the United States, 0.031 versus 0.082. Controlling for city size did not remove the difference: in every size class, the Canadian scores were substantially lower than the US ones. The index did not account for the many special-purpose district governments found in the United States, elected commissions overseeing a single service such as sewerage or waste management. Given that these quasi-governments are far more present in US metro areas, fragmentation there was even higher than reflected in Goldberg and Mercer's index.[1]

Donald Rothblatt found the trend persisting into the 1990s. He selected a sample of large metro areas in Canada and areas in the United States that were comparable in terms of their ranking in each country's metropolitan size distribution and calculated government density as the number of governments per million total metro population.[2] The government density on average in US city-regions was 31.8 compared with 17.8 in Canada. We updated Rothblatt's figures for 2010/11, with the results shown in table 5-1. The data show that although government density has changed very little in the Canadian sample (rising to 18.8), it has increased to 43.3 in the US group, indicating that

TABLE 5-1.
Governance characteristics, selected Canadian (2011) and US (2010) metro areas

Metro Area	Population (in thousands)	Area	Local Governments	Government Density	Central City Population	Metro Population in Central City (%)
Canada (2011)						
Toronto	5,583	2,280	24	4.3	2,615,060	46.8
Montreal	3,824	1,644	91	23.8	1,649,519	43.1
Vancouver	2,313	1,113	39	16.9	603,502	26.1
Edmonton	1,160	3,640	35	30.2	812,201	70.0
Average	3,220	2,169	47.3	18.8	1,420,071	46.5
United States (2010)						
Chicago	9,461	7,197	410	43.3	2,695,598	28.5
Boston	4,552	3,487	290	63.7	617,594	13.6
San Francisco (Bay Area)	7,468	6,907	110	14.7	805,235	10.8
Houston	5,947	8,827	160	26.9	2,099,451	35.3
Minneapolis–St. Paul	3,280	6,027	223	68.0	667,646	20.4
Average	6,142	6,489	238.6	43.3	1,377,105	21.7

Source: Statistics Canada Census; US Census Bureau.

fragmentation in the United States has progressed. Government density in the United States is now more than twice what it is in Canada. The table also shows the population weight of central cities in the metro area, again revealing significant differences between the two countries. The average of the cities in the sample of Canadian metros comprised almost half the metro population, whereas the average in the US sample made up just over a fifth of its metro area.

For those who take a free-market perspective on urban issues, the diversity of municipal governments in a given city-region is seen as a positive development. The competition among municipalities to attract residents and businesses creates a kind of marketplace, with each local jurisdiction offering its unique package of services and amenities. According to this so-called public choice view, intermunicipal competition ensures responsive administrations and efficient operations. Clearly, then, the more local jurisdictions, the greater the competition, and the more favorable the outcomes will be, other things being equal. On this view, there is no justification for state intervention to reduce the number of local municipalities (i.e., through annexation or consolidation). Local jurisdictions can collaborate on a voluntary basis to solve any regional problems that might arise.[3]

Because some residents and businesses undoubtedly do "shop" for a package of goods and services at a tax rate that suits them, intermunicipal competition may contribute to more efficiently run municipal services in some jurisdictions. The empirical evidence on this score is divided, however. On the one hand, several empirical studies have shown that fragmented regions appear to have lower per capita servicing costs, but other studies suggests that these lower costs may be due to lower servicing standards in such systems and not to greater efficiencies.[4] The tendency for servicing costs to rise following consolidation among previously independent jurisdictions is well known, but it is often due to higher and more equitable servicing standards throughout the expanded jurisdiction than in fragmented regions.[5]

Efficiency, however, is only one dimension of this issue. According to their critics, public choice advocates fail to see the larger problems related to municipal competition within regions. Local jurisdictions acting alone to manage what are essentially regional issues produce counterproductive results. They make decisions that often impose externalities on neighboring jurisdictions that may well be quite costly to the region as a whole.

The practice of fiscal zoning is a case in point. Because local governments finance their services largely from property tax revenues, they typically want to attract residents and businesses that will pay higher taxes and consume fewer services. Thus, each jurisdiction has an incentive to pass zoning ordinances and adopt other planning policies that exclude potential land uses that generate fewer taxes and require more services, in particular high-density housing. Similarly, each municipality seeks new businesses that will pay more tax revenues than the cost of the services they will require. This dynamic creates a pecking order among jurisdictions, with exclusive suburbs at the top offering high-quality services at tax rates that are low relative to incomes and with distressed inner suburbs and central cities at the bottom, stuck in a downward spiral of poor services, high social need, and insufficient tax revenues. If central cities try to raise taxes to finance a new social program, many taxpayers—

especially wealthy households and businesses—will decamp to a nearby suburban jurisdiction with lower taxes and better services. The empirical evidence shows that beggar-thy-neighbor competition in a fragmented metro region institutionalizes social inequality and spatial segregation.[6] The downloading by senior governments of responsibility for more programs to the local level has exacerbated this effect.[7]

The social polarization that results from urban fragmentation and intermunicipal competition reduces the sense of shared destiny between central cities and surrounding suburbs. Writing in 1995, Anthony Perl and John Pucher noted the difference between US and Canadian city-regions in this respect.[8] In particular, they highlighted the lack of cooperation between urban and suburban jurisdictions in the United States and how it exacerbates poverty within central cities and concentrates affluence beyond the city limits. In the United States, suburban public officials see their responsibility as creating a legal, fiscal, and physical wall between their jurisdiction and the urban center to secure their constituents' perceived economic and physical well-being. The antipathy between suburban and central-city jurisdictions in the United States helps explain why so many suburban municipalities have shown so little interest in institutions of regional governance or participating in mutually advantageous arrangement with central cities for delivery of cross-boundary services like transit. In Canadian metros, in contrast, Perl and Pucher saw evidence of what they call "reciprocal accessibility," in which suburbanites continued to see themselves as part of a larger region with a healthy central city and were therefore willing to support investment in regional transit infrastructure.[9]

In a fragmented metro area, autonomous municipalities acting alone are usually powerless to address broader issues with a regional dimension. That can be seen when suburban municipalities try to protect themselves from the rising tide of suburban sprawl by adopting local growth controls such as service area boundaries, building permit caps, very large lot size minima, frontage requirements, or bans on multifamily housing.[10] These controls not only raise housing prices and create barriers for the migration of lower-income households; they also work to push developers to the next community with more lenient development controls, resulting in a classic pattern of leapfrog development. Thus, municipal growth controls applied in a context of jurisdictional fragmentation may actually increase the rate of sprawl.[11]

Several research studies have demonstrated the link between government fragmentation and urban sprawl. After analyzing density trends the largest metropolitan areas in the United States between 1982 and 1997, William Fulton and his colleagues concluded that those areas with myriad small local governments sprawl more than those with larger units of local government (city, township, and county).[12] Similarly, Edward Glaeser and his colleagues found that employment location was more decentralized ("job sprawl") in more governmentally complex regions.[13] John Carruthers and Gudmundur Ulfarsson found that higher numbers of suburban municipalities and special districts correlate with lower metropolitan population densities.[14] In subsequent work, Carruthers found that governmental complexity was positively correlated with population decentralization as indicated by population growth in outlying unincorporated areas.[15]

Municipal Reorganization

In chapter 2, we discussed the degree of control that Canadian provinces have over municipalities compared with most US states. This difference arose in part due to the differential evolution of state-municipal powers in the two countries. In the colonial era, cities were governed as an extension of imperial rule. Chastened by the experience of the American Revolution, in which city governments played a role in organizing protest and resistance to the imperial power, the British ensured that cities in loyalist British North America would cleave to colonial governments, which gradually evolved into today's provincial administrations. Those governments retained full authority for "local affairs" under both the British North America Act (1867) and the Canadian Constitution that replaced it in 1982. As a result, the provinces have an undisputed authority to modify municipal institutions or create new ones without the consent of affected municipal government or local voters.

In the United States, many in the new republic saw autonomous cities as an expression of democratic aspirations and liberal freedoms. This approach worked well as long as most of the population was thinly settled in rural areas. In the nineteenth century, however, the lack of attention from the state government combined with enormous urban growth inevitably led to widespread corruption among local officials. In an attempt to assert the preeminence of the state over municipalities, Judge John Forrest Dillon enunciated his famous dictum in an 1868 case: "Municipal corporations owe their origin to, and derive their powers and rights wholly from, the legislature. It breathes into them the breath of life, without which they cannot exist. As it creates, so may it destroy. If it may destroy, it may abridge and control."[16] The judge did not have the final say, however: state intervention (such as forced annexations or amalgamations) into local affairs produced its own counterreaction, giving rise to the home rule movement. The result has been significant rollback of the power states exercise with respect to local governments, with many states (such as California, Georgia, Tennessee, and Arizona) exhibiting both Dillon and home rule characteristics.

The Canadian provinces have regularly exercised their power over municipalities to consolidate jurisdictions or create regional governance structures designed to manage growth and help finance strategic infrastructure. Among Canadian metropolitan regions, Toronto stands out for the number and variety of changes wrought by the Province of Ontario to its mode of governance in the post–World War II period. Best known was the 1954 creation of Metropolitan Toronto, what Canadian urban scholar Frances Frisken calls "the most sweeping change in city-region governance in North America since the New York State legislature consolidated 25 municipal units into NYC [New York City] in 1898."[17] Metro Toronto was a regional ("upper-tier") government imposed on the City of Toronto and twelve surrounding towns and townships without a referendum. In 1967, a round of mergers was conducted among the thirteen municipalities in Metro Toronto, resulting in a six-municipality configuration. In 1998, the province chose to completely merge the area into a single-tiered City of Toronto, notwithstanding an overwhelmingly negative vote in a referendum involving all six of the affected municipalities.[18]

The consolidation creating the new City of Toronto was part of a wave of restructurings in Ontario that nearly halved the number of municipalities in Ontario between 1995 and the early 2000s. Amalgamations and annexations have also been part of the provincial toolbox in Manitoba, Alberta, Quebec, and the Atlantic Provinces. British Columbia is the one Canadian province in which changes to municipal boundaries are usually decided by local residents through referenda instead of provincial government fiat, but there have been no recent examples in which significant changes have been approved.

In the United States, states have played a much more modest role in the reorganization of central cities and their regions. Historically, most amalgamations were of a voluntary nature, a process that proceeded smoothly as long as central cities were wealthy relative to surrounding rural and suburban areas. As wealthy suburbs began to develop in the second half of the nineteenth century, however, voluntary amalgamations become more rare. In the words of urban historian Kenneth Jackson: "The first really significant defeat for the consolidation movement came when Brookline spurned Boston in 1874.... After Brookline spurned Boston, virtually every other Eastern and Middle Western city was rebuffed by wealthy and independent suburbs—Chicago by Oak Park and Evanston, Rochester by Brighton."[19] Since World War II, there have been only a handful of voluntary consolidations of large central cities and their surrounding suburbs, such as those in Indianapolis, Nashville, and Louisville.[20] As for forced marriages of jurisdictions, no state in the United States has exercised its power to compel amalgamations in more than a century, and many of them have changed their laws or amended their constitutions to make it difficult to impossible for them to alter municipal boundaries or municipal institutions without the consent of municipal electorates.

Municipal Autonomy

The greater degree of provincial oversight in Canada compared with the level of state control in the United States is evident in policy areas other than municipal reorganization. In matters related to land use planning and regulation, Canadian cities experience substantially less autonomy than their US counterparts. Municipalities in Canada have access to the full range of tools such as zoning, planning, subdivision controls, and impact fees, but they are exercised under close provincial review. Most provinces require that municipalities over a certain population size (often 1,000) create a comprehensive plan. Local land use plans often have to be approved by provincial departments of municipal affairs, and development permits can only be issued after obtaining authorization from provincial environment departments. Provincial transportation, housing, public works, and economic development ministries play a variety of roles in reviewing and approving municipal decisions. Most provinces have planning policy frameworks designed to ensure the efficient use of land and public infrastructure, the provision of affordable housing, and preservation of farmland and greenspaces. Municipalities are required to respect these principles in their land use planning decisions and can be challenged in court or before a specialized tribunal if they do not.

A good example of this type of close review can be seen in the land budgeting process in the Ontario. Comprehensive community plans designate land for urban development, and development outside these areas is usually restricted. Municipalities negotiate the amount of land needed for development within the twenty-year time frame of the plan with provincial planners. The amount of land needed is calculated based on population and employment growth forecasts along with provincial density and intensification policies. The growth boundary is renegotiated during plan revisions every five years or so and can only be changed between plans if part of a comprehensive review process and approved by the province. Municipal planning decisions that violate the growth boundary can be appealed to a quasi-judicial tribunal.

The existence of these tribunals is another feature that distinguishes the Canadian planning landscape from what is typically seen in the United States. In Ontario, parties aggrieved by local planning or land use decisions, including inconsistency of any local bylaw or action with provincial plans or policies, can appeal to the provincially appointed Ontario Municipal Board (OMB). Established in 1897, the OMB is an administrative body with the power to overrule municipal actions, including appeals under the Planning Act related to matters such as zoning, subdivision plans, official plans, consents, and variances. Alberta, Saskatchewan, Manitoba, and Nova Scotia also have provincially appointed boards hearing appeals to municipal decisions, although their mandates are usually somewhat limited compared with that of Ontario.[21] These provincial appeal boards help ensure that municipal land use decisions are less arbitrary and self-serving than they might otherwise be and that they take into account wider principles of good planning.

Another feature of provincial-municipal relations that is of particular relevance to students of sprawl is the limited options open to Canadian municipalities in terms of the use of inducements to attract private investment.

Municipalities in Canada are not generally permitted by provincial legislation to offer any form of tax abatement, loan, loan guarantee, or grant to commercial or industrial entities.[22] This restriction limits the ability of local authorities within a city-region to compete with one another to entice high-tax-paying companies to locate within their borders. In contrast, most US states permit financial incentives to private-sector firms as location inducements. In the few US jurisdictions that have prohibitions on inducements, researchers have noted the advantages in terms of reducing intercity competition.[23] Indeed, John Stevens and Robert McGowan suggested that the states "should be the policy conduit for power and resources in local economic development, especially in dealing with economic development issues that cut across local jurisdictions."[24] Such is the dynamic found in many Canadian jurisdictions, where Canadian provinces tend to play a larger role in local economic development decisions, balancing the competing needs of constituent local governments—as well as those of private-sector bodies—and citizens.[25]

Some changes in municipal-provincial relations suggest that close provincial supervision may be slackening somewhat. Beginning in Alberta in 1994, a trend has developed toward increasing the autonomy of municipal governments. Instead of listing a set of narrow functions, the Alberta Municipal Government Act laid out general spheres of jurisdiction within which municipalities could operate and attributed broad corporate powers (e.g., to purchase property, enter into public-private partnership agreements, sue for breach of contract) to municipalities. These changes have been echoed in British Columbia, Quebec, Manitoba, Ontario, Saskatchewan, and Nova Scotia. Another shift is the allocation of greater powers to the major cities in Canada. Although some cities, like St. John's in Newfoundland, have always had city charters that enhanced their powers relative to other municipalities in their respective provinces, more cities are gaining this extra jurisdictional latitude. Now, Vancouver, Montreal, Saskatoon, Regina, and Toronto all have charters, which typically afford greater latitude to city councils in terms of fiscal instruments and policy innovation.

The fiscal relationship between the provinces and municipalities has also evolved in a way that gives the latter more autonomy with respect to other levels of government within the federation. Since 1990, there has been a move away from the use of conditional grants in many provinces, which reduces the leverage the provinces have over municipal decision making. Indeed, total transfers from senior governments fell significantly in the 1990s while program responsibilities downloaded onto municipalities increased, thus increasing pressure on municipalities to find alternate sources of funding. Among other strategies, many municipalities turned to imposing impact fees, or levies on private developers to fund municipal infrastructure needed to support growth. The increased proportion of municipal revenue from own-source revenues has increased municipal autonomy, but it has done so in a way—at least potentially—that can help manage fringe growth.[26]

Traditionally, infrastructure associated with new development was funded through general taxation, which spread out costs onto all taxpayers and artificially stimulated development activity. By resorting to impact fees, municipalities forced developers to internalize some of the costs of growth and brought more order to fringe development. Charges as high as $75,000 per dwelling unit are not unheard of in Ontario, and many other provinces also have hefty levies on development. The charges may be applied on a per-unit or per-frontage-foot basis and work to increase densities as developers attempt to reduce the effect on selling prices. To encourage intensification, many cities reduce or eliminate impact fees in already established areas. In some cases, charges are structured on an area basis to reflect parameters such as the distance from treatment plants or other facilities.[27] Canadian cities tend to rely on such charges to a much larger extent than US cities, which may account for some of the difference we observed in urban form.

Metropolitan Governance

The debate on the desirability of regional governance institutions is a long-lasting feature of both Canadian and US urban policy discussions. Since at least the 1920s, when Clarence Stein and his colleagues formed the Regional Planning Association of America, progressive voices have been arguing for formal institutions that would knit together central cities with the burgeoning suburbs taking shape beyond their borders. Traditionally, advocates of regional government have highlighted the importance of maximizing economies of scale in service delivery, managing urban growth, planning major infrastructure improvements, and reducing social disparities. In the 1960s, however, public choice advocates began challenging this view by making the case that services were more effectively delivered at the local level, where accountability and transparency were highest.[28] In the 1990s, globalization and the increasing attention to metropolitan areas in the battle for economic dominance brought increasing pressure on finding ways to address regional needs and attend to issues that transcend local political boundaries. One result is the new regionalism movement, which emphasizes the potential of voluntary cooperation among municipal officials, business groups, and other stakeholders over the creation of new layers of government.[29]

In its efforts to manage metropolitan growth, Canada has shown a clear preference for solutions that emphasize the reorganization of regional government structures and responsibilities. This approach draws on the nature of Canadian federalism, which puts the responsibility for urban affairs in the hands of the provincial governments and provides those governments with the power to create, adjust, and eliminate municipal entities, in an almost unfettered way. The evolution of the Canadian urban governance system also gives provinces the power to oversee municipal planning and create regimes that ensure the transmission of provincial policies into municipal planning decisions affecting development on the ground.

In contrast, the United States, with a few exceptions, has favored the decentralized collaborative approach. The most important exceptions are Portland, Oregon, and Minneapolis–St. Paul; these metropolitan areas use a three-tiered regional system of governance, with metropolitan councils empowered with regional planning authority over a geographical area that incorporates several counties and their component cities. Miami-Dade's two-tier system avoided the creation of a new layer of government by assigning more powers, including county-wide land use and transit planning to the existing country level. A few dozen county-city consolidations have created larger municipal units in some cases, but they rarely account for more than half of the regional population and usually much less.

Beyond these exceptions, most regional mechanisms in the United States are based on narrowly focused service delivery agencies (e.g., transportation, waste, sewer, education) or voluntary arrangements among local authorities. The latter include councils of government in places like metropolitan

Washington, DC; Baltimore; the San Francisco Bay Area; and Southern California. These councils are weak organizations with no taxation or legislative authority, with member governments being entitled to ignore the advice proffered, whether on land use, transportation, or housing issues. They provide a metropolitan forum where their members can discuss various issues, opportunities, or common problems. They have accomplished little in the way of managing growth or stemming sprawl, however, and they do not function to redistribute public benefits from the haves to have-nots.[30]

Despite the shift toward greater collaboration in the United States, doubts persist that decentralized governance solutions will be able to generate the sustained policy focus, integrated vision, and leverage to resolve regional governance problems. Research shows that regional structures and state involvement in planning matters—where it exists in the United States—have helped reduce sprawl and push development toward already urbanized areas.[31] Calls for institutional reform are still heard in the United States, despite the acknowledged political obstacles that such reforms would have to scale. In any difficult situation, one of the main obstacles to change is the tendency for people to favor the status quo, the devil they know. Our perception of choice is bounded by our traditions, economic practices, social structures, and culture. When it comes to institutional reforms, Canada has a much wider palette than is available in the United States; in fact, Canada is a world leader in experimenting with a wide variety of metropolitan governance models. Let us turn now to an exploration of some of these options as potential guides for the United States in this twenty-first century.

Single-Tier Systems

Several of Canada's metro regions are governed by a central city that comprises the vast majority of the metropolitan population. In these cases, the central municipality effectively functions as a form of single-tiered metropolitan government. "Unicities" can result from gradual annexations or consolidations (called amalgamations in Canada). The annexation approach is especially popular in the Alberta and Saskatchewan, where the cities of Calgary, Regina, and Saskatoon all comprise more than 85 percent of their respective regional

TABLE 5-2.
City and metro populations, selected Canadian unicities, 2011

City	City Population	Metro Population	City Population of Metro Population (%)
Calgary, Alberta	1,096,833	1,214,839	90
Regina, Saskatchewan	193,100	210,556	92
Saskatoon, Saskatchewan	222,189	260,600	85
Winnipeg, Manitoba	663,617	730,018	91
Sudbury, Ontario	160,274	160,770	100
Halifax, Nova Scotia	414,400	408,702	100

Source: Statistics Canada Census.

populations (see table 5-2). To the east, unicities have largely resulted from the consolidation of neighboring municipalities into a single city (Halifax) or the collapse of a two-tiered system into a single tier (Winnipeg, Sudbury).

The city of Calgary's territory has increased incrementally since its founding in 1884 as a result of more than forty annexations of surrounding rural areas (figure 5-1). The city annexes territory in the path of its expansion so as to maintain a thirty-year supply of developable land. Provincial policy prevents rural municipalities from permitting incompatible development in the boundary zone with Calgary so as not to complicate the city's expansion plans. The annexations are part of the city's sophisticated growth management system, comprised of long-term strategic planning that links growth forecasts to the city's district plans and rolling capital budgets.[32] Since the 1990s, rising traffic congestion and infrastructure costs have pushed the city to use its planning machinery to increase densities and focus growth around transit stations. The 2009 master plan requires 30 percent of growth to be accommodated through intensification and for greenfield developments to achieve a target density of twenty-four people plus jobs per gross acre. Canadians tend to think of Calgary as "sprawl city," but its close-knit suburban layouts and lack of leapfrog development might appear unusual in a US context.

Although the unicity approach has proven effective as a means of efficiently managing growth, it is not without its problems. Boundary changes are often unpopular with residents in the targeted areas, which results in drawn-out

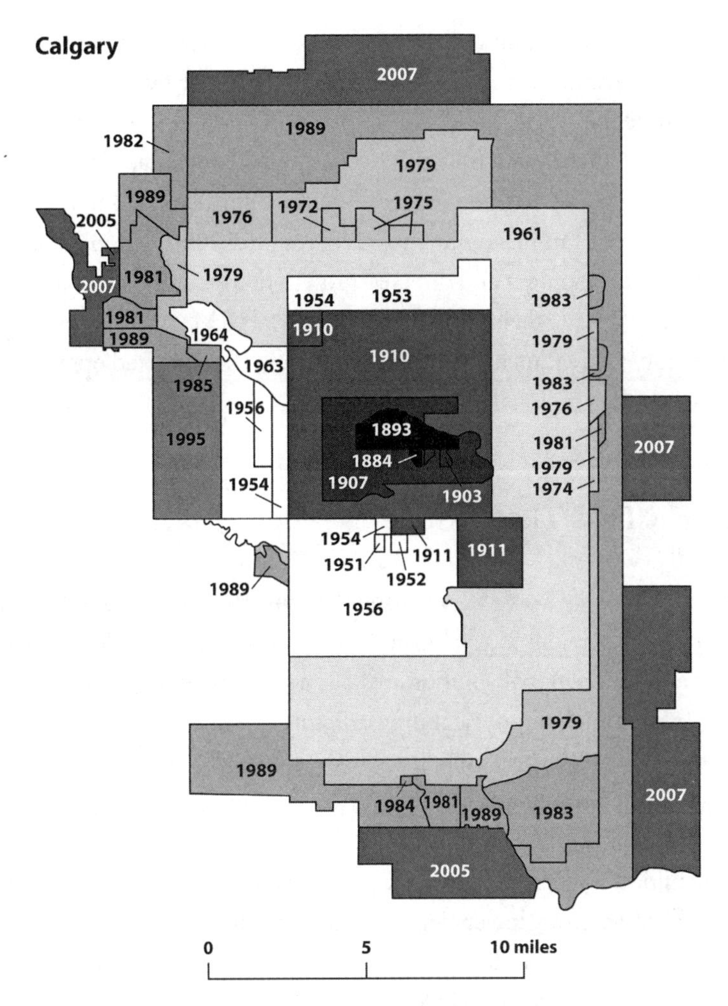

FIGURE 5-1. Geographical growth of Calgary through annexation, 1883 to present
Source: Taylor, Burchfield, and Kramer 2014.

disputes and damaged relations among neighboring municipalities.[33] Another issue is that annexations cannot keep up with the accelerating growth in towns and rural municipalities in the wider region, creating a demand for some form of metro-wide coordination of land use and transportation issues.

The Halifax Regional Municipality (HRM) was created through a 1995 provincial law that merged three single-tiered urban municipalities and one large rural county. Prior to that, the region had hobbled along with an ineffectual regional planning authority, but the 1995 merger made this body and its plan irrelevant. Consolidation was brought about by a desire to save money on municipal services and to eliminate intermunicipal competition for economic development, which was undermining the regional plan. There is no evidence that money has been saved,[34] but the amalgamated municipality has made major steps in managing growth through a 2006 regional plan that restricted rural development and directed growth to the central areas and transit-supportive nodes. A revised plan, adopted by HRM in 2014, committed the municipality to creating a greenbelt, or a system of connected open spaces, to further strengthen regional planning goals.

Top-Down Two-Tier Systems

The basic principle of two-tiered metropolitan government is straightforward: an upper-tier government is established for those functions of local government that require a regional solution, usually region-wide land use planning and major intermunicipal physical infrastructure. Because taxation is usually based on each municipality's property assessment, regional spending is an effective way to redistribute taxes within a two-tier federation. Lower-tier municipalities remain in place to provide local services such as zoning and recreational facilities. Typically, upper-tier governments approve major planning documents coming from lower-tier governments, such as their community plans, secondary (district) plans, or subdivision plans. The upper-tier plans are usually approved by the province.

The Greater Toronto Area (GTA) has seen the most extensive use of two-tiered regional government in Canada. After World War II, Toronto experienced very rapid growth and by the early 1970s had bested Montreal as Canada's largest metropolitan area. The rapid rate of suburbanization outside the borders of the central city was overwhelming the capacity of those municipalities to cope with the needed infrastructure. This growth prompted the province to pass legislation in 1953 that created Metropolitan Toronto, a two-tiered structure that covered 90 percent of the metropolitan population incorporating Toronto and twelve surrounding municipalities, later reduced to five (figure 5-2). The upper-tier council was indirectly elected and had no powers of

(a)

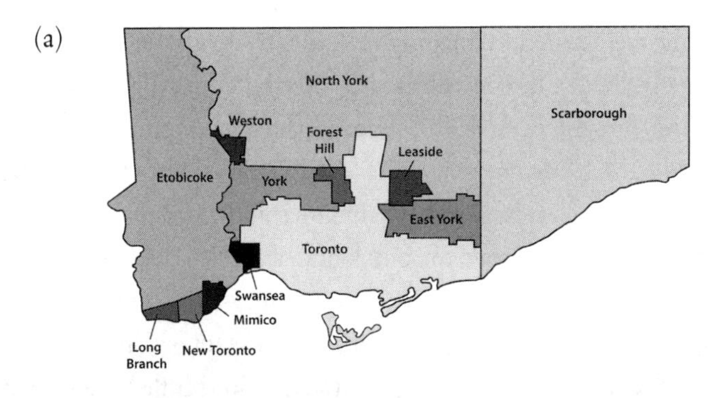

(b)

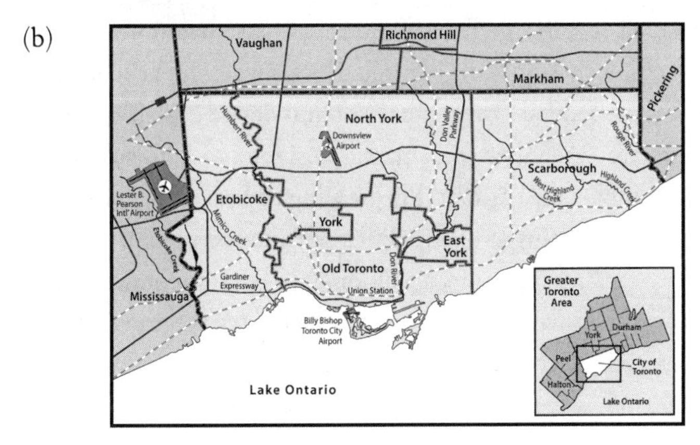

FIGURE 5-2. Metropolitan Toronto and component municipalities: (a) 1953, (b) 1967
Sources: (a) City of Toronto; (b) Lencer on Wikipedia.

taxation; its revenues came from the member municipalities in proportion to their assessment bases. The arrangement was widely hailed as a successful compromise between local autonomy and regional coordination. The large commercial tax base of the central city subsidized infrastructure and growth in the surrounding suburbs in return for higher suburban densities and an emphasis on building transit infrastructure to bring workers downtown. Two-tiered governments more or less modeled on the Toronto solution were set up elsewhere in Ontario as well as in Winnipeg (Manitoba) and throughout southern Quebec, including Montreal.

Metro Toronto worked relatively well for more than thirty years, but tensions grew in the 1980s when the suburban municipalities balked at the cost of refurbishing infrastructure in the aging central city. As upper-tier expenditures grew, pressure to have a directly elected upper-tier council mounted, and this change was put into place in 1988. That move only aggravated tensions on Metro Council, which became somewhat dysfunctional after that. In 1998, the province eliminated the lower-tier municipalities and created a new City of Toronto, citing the need to improve the efficiency of service delivery, better attract private investment, and—perhaps most importantly—handle the extra costs imposed on it due to the downloading of services such as transit and social services. The available evidence suggests amalgamation has done little to save on costs.[35]

Ontario created ten other two-tiered regional governments in the 1970s. The largest, Ottawa-Carleton, was relatively successful at managing growth. In the 1950s, the federal government commissioned a plan that proposed the creation of a large greenbelt for Ottawa, designed to hem suburban development into a compact area around the growing central city. The greenbelt came to be, but suburbanizing municipalities outside the greenbelt continued authorizing low-density bedroom development. To gain control of this leapfrog development, the provincial government created the Regional Municipality of Ottawa-Carleton (RMOC) in 1969, an upper-tier government that brought together sixteen lower-tier suburban and rural municipalities within its extensive boundaries. The first regional plan, approved in 1974, was based on concentrated development in a few areas outside the greenbelt and good-quality transit to reduce automobile dependency.

In the 1990s, the RMOC moved from indirect to direct election of upper-tier councilors, a move that increased bickering between the two levels of government. The provincial government, intent on reducing the burden of government on citizens and the private sector, took this opportunity to abolish the lower-tier municipalities and consolidate the two-tiered system into a one single-tiered city.[36] The new city soon after adopted a smart growth plan that would curtail low-density growth in the fringe and move toward stronger transit-oriented forms of development. For example, the plan set a target of twelve dwelling units per acre in the suburban areas outside the greenbelt, aimed for 36 percent of new growth to be accommodated through intensification, and attempted to strengthen growth nodes by steering new residential and employment growth toward them. Despite some resistance from land developers rural interests on city council, the plan has been successfully implemented in key respects.[37]

Bottom-Up Two-Tiered Systems

In another approach to regional governance in Canada, the province sets the rules of the game and then leaves municipal partners in the region to work out their level of collaboration and develop a regional plan. This approach was attempted in the Edmonton metropolitan area and is currently being used in metropolitan Calgary. Both these metros were left without effective governance when their regional planning commissions were abolished by the province in 1995 due to conflict with member municipalities. In their place, partnerships were created to address regional planning issues, but according to the provincial legislation that defined the partnerships, municipal membership was not mandatory. The Edmonton partnership was crippled when the central city withdrew, forcing the province to create a new top-down planning board. In Calgary, the partnership continues, but the regional plan that emerged is being undermined by the refusal of some rural districts to participate.

British Columbia uses a more structured bottom-up two-tiered system, one that has been highly successful in places like Vancouver. Metro Vancouver (formerly the Greater Vancouver Regional District, or GVRD) is a regional district that includes twenty-one municipal partners and covers virtually the entire metropolitan area (figure 5-3). Officials are careful not to refer to Metro Vancouver as a "government" because the term implies a level of control that the regional district does not have. Rather, decision making by the district board (council) is based largely on voluntary cooperation among member municipalities within a legislative framework established by the province, an approach that has been called a "strategy of gentle imposition."[38] Originally charged by the province with regional planning, potable water, and sewage mandates, the board decides by consensus what additional services will be delivered through the regional agency. Municipalities that sign on for those services are billed accordingly, without any direct taxation. Over time, the board has taken on regional parks, waste management, public housing, and air quality.

Since adopting its first regional plan in 1975, Metro Vancouver has had a consistent planning vision for the region based on improved transit and reduced car use, with growth concentrated in the metropolitan core and town

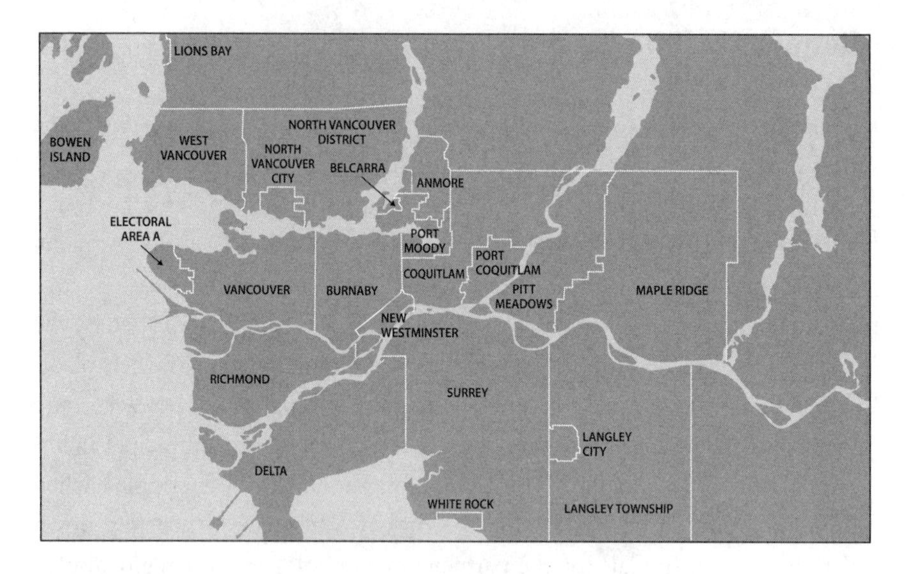

FIGURE 5-3. Metro Vancouver

centers, along with a regional greenbelt. The greenbelt is largely made up of farmland that has been protected from development under provincial law since the 1970s. After an interregnum during which its regional planning powers were abolished by the province due to conflict with member municipalities, the board's planning powers were restored in 1995, and an updated plan was adopted in 1996 and then again in 2011. Under the province's *Growth Strategies Act*, regional plans can only be adopted by consensus of the district's board, which comprises representatives from member municipalities. There is often contentious debate among members of the board, and decisions are excruciatingly slow, as evidenced by the ten-year time line on the most recent regional plan. The system has proved to be resilient, however, capable of addressing regional policy issues and providing efficient regional services.[39] Vancouver's very high intensification rate (75 percent of all growth is in already built-up areas), overall density, and high transit ridership are testaments to the success of its planning approach. Less flattering are the sky-high housing prices, which many also attribute to Metro Vancouver's regional planning efforts.

Multiple Single- and Two-Tiered Systems

Since World War II, metropolitan governance in the two largest metro regions in Canada, Montreal and Toronto, has struggled to keep up with the growth and spread of their respective regions. The two-tiered metropolitan structures developed in the distant past became overwhelmed by growth beyond their borders, leaving both regions with barely constrained fringe growth. This dynamic triggered intense debate in both regions over the proper structure of metropolitan governance, with the two cities going in somewhat different directions.

The consolidation of Metro Toronto in 1998 mentioned earlier encompassed barely half of the metro population. Back in the 1970s, the province had organized suburban areas outside Toronto into two-tiered governments to bring order to their explosive growth. The upper-tier governments were given responsibility for managing growth, but in the absence of strong urban centers with an interest in stemming sprawl and boosting transit, there was little progress on this front. Although there was little leapfrog development in these outer regions, most of this growth was at densities too low to support transit, typically at about seven units per gross acre.

In the absence of a metropolitan plan, the province tried to use consultation with municipalities to address regional growth issues throughout the 1990s. These initiatives helped build consensus around an informal "nodes and corridors" plan for the GTA, a plan that both municipal and provincial planning agencies were supposed to respect. Although it may have tempered sprawl a little, this effort was having little effect on the compounding problems in the region, including road congestion, inadequate transit, disappearance of farmland, and a ballooning infrastructure debt.

Between 1995 and 2002, the province turned for advice on regional governance to community leaders by setting up a series of advisory task forces. One task force recommended that the five existing upper-tier governments in the GTA be abolished and replaced by a new Greater Toronto Council, but due to resistance from suburban municipalities, this solution was rejected outright. A new consultative initiative was launched in 2002 when the province created a smart growth panel and tasked it with proposing solutions for the geographical area that became known as the Greater Golden Horseshoe (GGH). With a population of almost 9 million, the GGH covers the GTA along with the city of Hamilton and neighboring areas in a vast territory around the western end of Lake Ontario.[40] Instead of concentrating on institutional reforms, the panel focused its attention on planning issues. Echoing the earlier informal plan for the GTA, it proposed a system of concentrated subcenters connected by good-quality transit along with new highways to relieve congestion.

In 2003, the province was finally ready to take decisive steps. First, it established one of the world's largest greenbelts, a 200-mile-long band of farms, forests, and wetlands that runs through the center of the GGH.[41] Next, the province adopted Places to Grow, a land use and transportation plan for the GGH. Drawing heavily from the smart growth panel's report, the plan promotes growth in existing urban centers; encourages intensification and compact, mixed-use development; and projects significant new transit lines and highways (figure 5-4). The new plan allocates forecasted population growth to upper-tier municipalities, limits urban expansion areas, and lays down density (twenty people plus jobs per acre in greenfield areas) and intensification targets (an average of 40 percent) to guide growth in the region.

The greenbelt and Places to Grow have been controversial within the region. Some municipalities, farmers, and developers are disgruntled due to the reduction in development potential in some areas and imposition of planning targets. Densities and intensification rates are rising in the region, but that could simply be the continuation of trends already detectable prior to the provincial planning effort. Provincial officials are working with municipal planners to ensure that municipal plans faithfully reflect the growth management vision behind the provincial plan, but many of these municipal plans are being appealed to the Ontario Municipal Board. If a provincial election brings in the free-market Conservatives, the government may scale back its grand vision for managing growth in the GGH.

The Montreal story parallels that of Toronto in some respects. In 1969, the provincial government established a two-tiered municipal system on the Island of Montreal to provide regional services and coordinate infrastructure planning among twenty-nine separate municipalities. Even when it was created, this organization (called the Montreal Urban Community, or MUC) covered

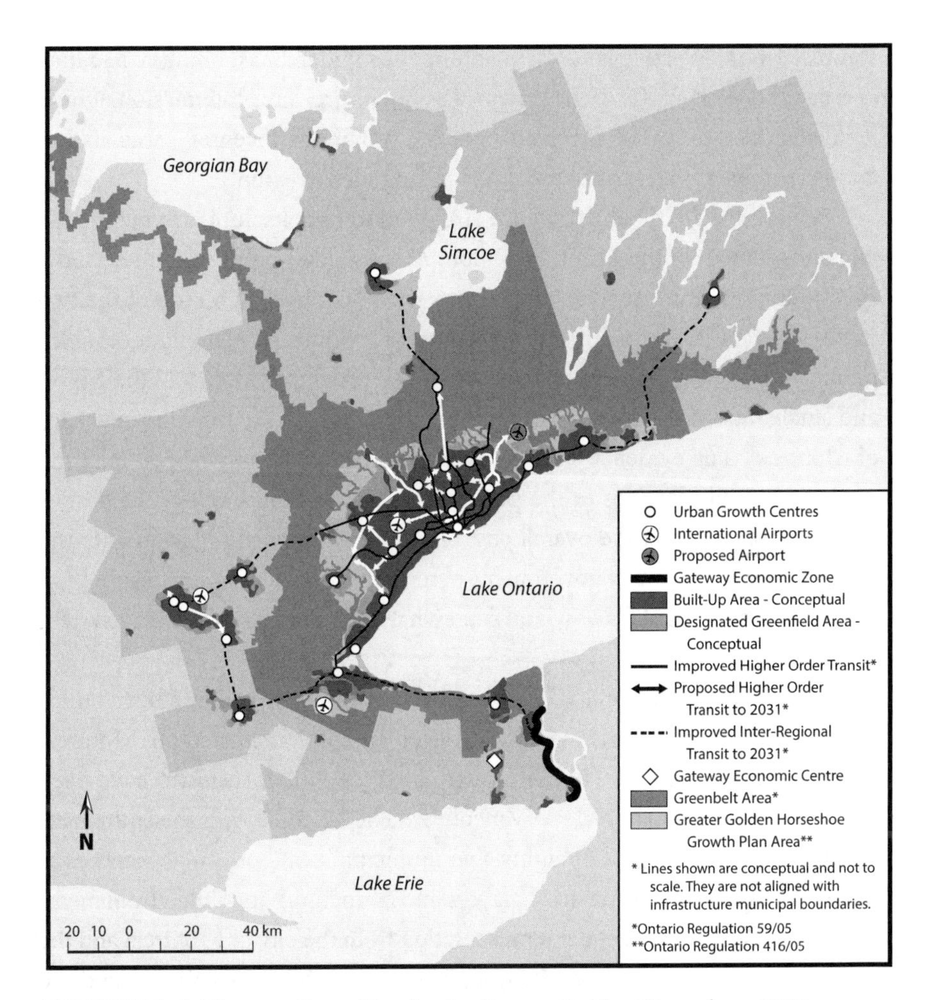

FIGURE 5-4. Places to Grow Plan for the Greater Golden Horseshoe, 2006
Source: Ontario Ministry of Infrastructure 2006.

only 70 percent of the population of the metropolitan area. The governing council was made up of representatives from the component municipalities, and because of city-suburban differences, the upper tier was unable to develop into an effective government.

Outside the MUC, more than one hundred off-island municipalities created a fragmented patchwork of local government that did not lend itself to effective growth management. As in Toronto, two-tiered municipal jurisdictions were introduced by the province to help coordinate services and stimulate more effective land use planning over wider areas, but they had limited beneficial effect. Only the province's agricultural land commission, which regulates the removal of farmland from the province's inventory, was able to temper sprawl on the fringes of the expanding metro region.

Through the 1990s, the province struggled to find a solution to metropolitan management in the Montreal region. Among other proposals, it rejected a special commission's scheme for a metro-wide government on top of the two existing tiers. Finally, in 2002, despite extensive opposition among the Anglophone and wealthy Francophone suburbs, the province eliminated the MUC and amalgamated the island's twenty-nine municipalities to form the new City of Montreal. The evidence suggests that the merger probably helped raise the quality of public services in the poorer areas of the island, but at the cost of pushing up taxes and the overall cost of municipal services.[42] Soon after the merger, a new provincial government permitted some demergers, resulting in the reestablishment of a messy and somewhat dysfunctional two-tiered system on the island.

Although it attracted much less attention, the province had also been building a new institution at the metro-wide level. In 2000, it created the Montreal Metropolitan Community (MMC), with a structure that seems to have been inspired in part by that of Metro Vancouver. The MMC covers the entire metropolitan area, comprised of eighty-two municipalities, both single- and two-tiered. The council is made up of mayors and councillors appointed from member municipalities, with equal representation from the city of Montreal and the surrounding suburbs. The MMC was given responsibility for regional land use planning, economic development, social housing, regional infrastructure, and solid waste management. The council has no power to tax residents or businesses, but levies member municipalities to finance its activities. As in Vancouver, the metropolitan authority has no direct responsibility for transit planning.

In 2009, after a false start when a draft metropolitan plan sat unapproved by the MMC council for several years due to suburban resistance, the province ordered the decision makers to produce a plan within a two-year time frame. With this impetus, a plan for the entire metropolis finally emerged in 2011. The plan imposes a growth boundary, requires that 40 percent of development in the region take place around transit stations (figure 5-5), and obliges outlying municipalities to achieve minimum greenfield densities of about ten units per

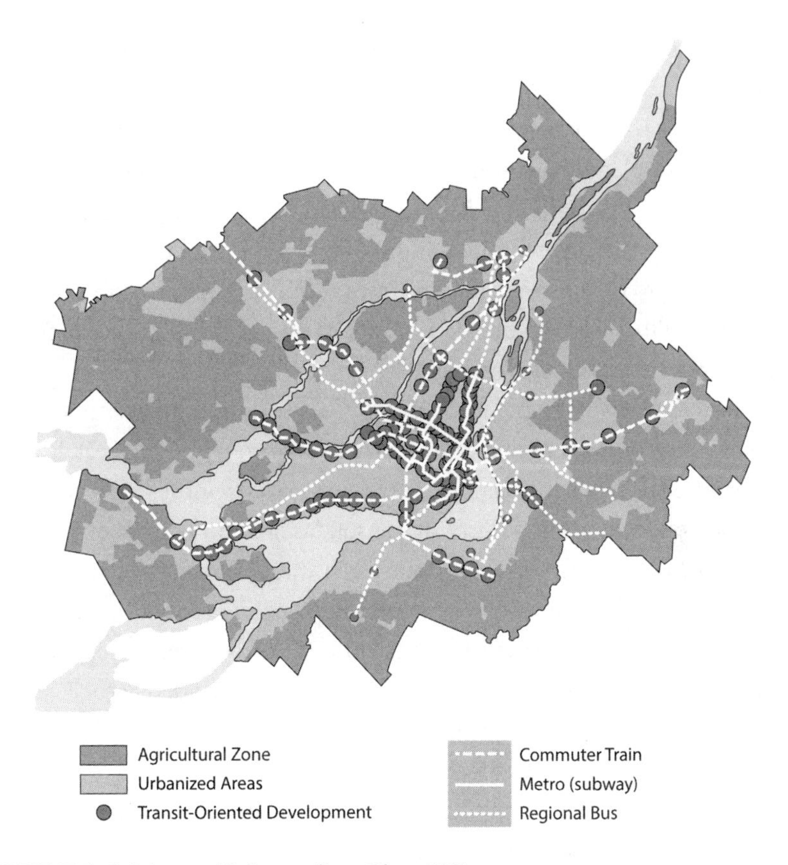

FIGURE 5-5. Montreal Metropolitan Plan, 2011
Source: Montreal Metropolitan Community, 2011.

acre by 2031 (up from the current seven units per acre). The plan also sets a 35 percent modal share for transit by 2031, up from the current 20 percent share. MMC planners are now working with municipalities in the region to bring their community plans into conformance with the new regional plan.

Conclusion

Canadian scholars who took an interest in issues related to metropolitan governance in Canada during the 1980s and early 1990s noted with disappointment that provincial governments there appeared to have stepped back from their commitment to creating structures that could provide strong regional management. They noted that Alberta had abolished its regional planning commissions, the GVRD and other regional districts in British Columbia had had their statutory planning powers abolished, and the two largest metros areas in the country—Toronto and Montreal—had been carved up into multiple two-tiered jurisdictions with no metropolitan plan in sight. They feared that Canadian metro regions were moving toward US-style governance as antigovernment, neoliberal ideas spread into Canada and growing suburban municipalities exercised their political clout.[43]

From today's vantage point, however, we see little evidence that the Canadian system of metropolitan governance has drifted into a US-style governance approach. Instead, there has been a resurgence of regional restructuring initiatives, including a flurry of consolidations, further annexations, the creation of metropolitan councils, and even direct provincial planning on a supermetropolitan scale. Moreover, there is no evidence that provinces have abandoned their commitment to invest in and support central cities, which remain relatively healthy across Canada. Finally, no tendency for Canadian cities to incorporate private or nonprofit interests directly into the planning process has emerged. With few exceptions, Canadian students of metropolitan governance persevere in the opinion that some form of institutional centralization is needed to manage regional growth effectively.

The increasing attention given to metropolitan issues in Canada is driven by issues similar to those driving interest in this subject in the United States. After a couple of decades of relative inaction, the metropolitan governance arrangements in both countries started to show their cracks in the 1990s as congestion worsened, the infrastructure debt ballooned, and the environmental costs associated with sprawl and car dependence came into focus. The failings of the status quo, in other words, became visible not only from the vantage point of distressed central cities concerned for their viability within existing metropolitan arrangements, but also from the point of view of the suburbs themselves. The difference between the two countries is that while the United States turned to greater reliance on intermunicipal agreements and private-public partnerships, Canada has chosen to go further down its path of provincially directed institutional innovation and direct involvement in local planning.

The US approach is characterized by the relative absence or weakness of state and regional influence on local decision making. In the United States, most urban areas contain dozens of autonomous local governments, with each individual municipality adopting its own master plan and zoning ordinances and little in the way of regional or statewide frameworks or concerns to constrain it. The metropolitan coordinating structures that have been set up have had some success where goals are clear-cut and limited, but they have not proven effective at regional governance, especially in fulfilling the functions of most concern to us, namely managing growth and shaping urban form to stem sprawl and support a regional transportation system that will limit the attractiveness of the private auto. Those are the thorny issues that voluntary regional associations tend to avoid precisely because they threaten the autonomy of local jurisdictions.[44]

In Canada, not only does there tend to be a smaller number of municipalities within a given metropolitan region, but the regional governance systems that prevail there impose a degree of constraint on the autonomy of local jurisdictions with respect to land use. As opposed to the overlapping authority typically found in US metro regions, Canadian metros are characterized by what political scientists call coordinated authority.[45] Local land use planning in most parts of Canada takes place in what one can characterize as a culture of accountability, where individual municipalities answer to varying degrees to regional and provincial bodies and must act in ways consistent with regional or provincial plans and policies. Although there is no guarantee that larger bodies will be more sensitive to broader growth management and environmental concerns—for example, the OMB has been criticized for being too growth- and developer-oriented in its decisions—they are likely to address the larger issues that are often not recognized at the local level. A framework in which infrastructure investment strategies are decided at the regional level and where local planning is accountable to regional or provincial-level bodies, and must operate within parameters set by those bodies, is significantly more likely to foster growth patterns that are sensitive to larger transportation, energy use, and environmental considerations.

The higher densities and less prevalent exurban growth found in Canada compared with the United States suggest that the more coherent, top-down planning regimes found in Canada have contributed to the relative lack of leapfrog development and the containment of urban sprawl. Urban growth boundaries—that is, regulatory provisions that make it more difficult for municipalities to expand the urban area beyond a certain limit—are relatively common in Canada. Less fragmentation compared to the proliferation of local authorities found in US metro areas has also helped reduce destructive competition among municipalities and dampened the forces of urban sprawl. The provincially established agricultural land reserves also limit growth in some provinces, such as British Columbia and Quebec. Set up in the 1970s, these reserves cover the farmland around the cities and prevent municipalities from converting viable farmland to urban uses without the approval of a provincially appointed commission.[46]

Another outcome of note is the tendency for Canadian metropolitan regions to have a more defined urban structure, with a strong center and subcenters linked by good-quality transit. Greater Vancouver is probably the best-known example along these lines, having benefited from several decades of a consistent regional planning vision as expressed in its growth management plans, starting in 1975 and reiterated in the 1996 and 2011. Ottawa is well known for its planned employment clusters around rapid transit stations, and Toronto is also known for linking urban planning to transit provision, producing a system of higher-density nodes and boosting transit use (discussed more in chapter 6).

The governance models used in many Canadian provinces have also contributed to greater social equity in Canada cities and city regions. Fiscal redistribution has worked both ways in Canadian metros, with well-off central cities sometimes subsidizing public services and infrastructure in adjacent suburbs and at other times with suburbs helping pay for expensive regional services in the central city. Full consolidation may not be necessary to achieve these benefits; experience in Canada showed that many of the equity benefits of municipal reorganization could be achieved with two-tiered government. Full consolidation, usually motivated by a desire to reduce administrative and serving costs and spur business investment, rarely met its promise.

That metropolitan government has had some success is not to say that the Canadian approach is without its stresses and strains. Because the purpose of metropolitan governance is to link central cities with their growing suburbs in a decision-making matrix, it is not surprising that Canadian metro areas are often marked by serious antagonism between urban and suburban interests. In many cases, the institutional arrangements that have been put in place in Canada have been able to contain, mediate, or override these tensions, bringing the city and suburbs together in a common bond instead of the sharp divisions one observes in many US regions. In other cases, regional bodies in Canada have been paralyzed by intermunicipal infighting, and in extreme cases, the friction has led to their complete abolition.

A related issue is a question often heard in discussions about metropolitan governance in Canada: who speaks for the region and the regional interest? In principle, the tiered governance system is supposed to provide a regional voice by constituting a body responsible for the entire region. In practice, however, indirectly elected councillors rarely dare espouse the regional point of view when it clashes with the wishes of their constituents. In a few cases, indirectly elected councils were transformed (by provincial order) into directly elected ones in an attempt to provide more accountability at the regional level. This strategy seems to have backfired; in most cases, it led to greater competition for legitimacy between the two levels of government. All told, the indirect system seems to have allowed upper-tier councils to muddle through on the basis of strong staff support and the tendency for good ideas to survive and win out in the end.

We have seen that the Canadian system of metropolitan governance relies on a few underlying processes: annexation, creation of two-tiered municipalities, consolidation of previously two-tiered jurisdictions, mergers between single-tiered municipalities, creation of limited-purpose metropolitan authorities, and the development of provincially led metropolitan plans. Which of these processes is of most relevance to the US situation?

City-county consolidation or annexation may continue to be a solution for smaller urban regions, but for major metropolitan areas that have long outgrown their county boundaries, this solution does not appear to be viable. In larger multicountry metro areas, some kind of multitiered system might be warranted. A top-down multitiered approach is not a politically viable option in most US metropolitan areas. As Bernard Ross and Myron Levine put it: "Suburbanites and local officials oppose major political restructuring: suburbanites are unwilling to surrender local autonomy for the vague promise of cost savings. Public choice theorists doubt the merits of metropolitan restructuring.... Racial minorities often object that metropolitan reform will effectively take power away from them just as their numbers have grown sufficiently large to gain electoral control of the central city."[47]

In his recent comparative review of US and Canadian metropolitan governance systems, David Hamilton concluded that a bottom-up tiered approach like that used in Vancouver was the most promising model for the United States. "With a tiered approach," he writes, "services can be organized and assigned to the appropriate level to maximize efficiency. All residents in the region can have equal opportunity to share in the quality-of-life value and have equal access to the benefits from economic growth. Finally, citizen participation is not diminished from a more centralized governing system and a political institution is in place to address regional policy and governing issues."[48]

Perhaps the collaborative approaches currently favored by new regionalists in the US will be able to solve the compounding challenges associated with urban sprawl, regional inequality, central-city decline, and car dependency. Or perhaps they will not. If not, these problems, which form a mutually reinforcing syndrome, could reach crisis proportions due to social unrest, resource depletion, or fiscal exhaustion. Under such conditions, institutional reform may once again get on the US urban agenda.

Chapter 6

Urban Connectivity: Integrated Transportation Planning

From a functional perspective, transportation networks and land use patterns are intimately related: transportation infrastructure moves people and goods between destinations that are separated or grouped in space as defined by the land use system. Land use decisions generate the need for transportation services, whereas investments in the transportation system stimulate development in specific locations. Transportation and land use planning need to work in tandem to achieve region-wide goals, such as the efficient use of existing infrastructure, vibrant city centers and subcenters, and optimum accessibility to goods and services.

Although planners have been aware of this imperative for a long time, integrated land use and transportation planning has eluded most jurisdictions in the United States. The major components of the urban transportation system are planned and developed on a large geographic scale, mostly funded by senior governments with their own agendas, whereas land development is managed on a microlevel by local planning authorities. The result is that land use is often inadvertently shaped by large-scale transportation investments, which are focused on immediate transportation objectives and not on longer-term land development goals. Meanwhile, large-scale infrastructure investments not supported by appropriate land use policies may fail to trigger the intended development and fall short on ridership levels and performance.[1]

Within local governments, transportation engineers and land use planners have tended to work in their separate silos with minimal interaction. Transportation engineers are usually solely responsible for providing easy and secure traffic flow on streets, in other words, ensuring good mobility in the street network. Anticipating and resolving traffic congestion are their primary concerns, with supply-side solutions such as expanding the road system usually the only ones on offer. Land use planners are usually more attuned to the larger picture than transportation engineers but have little control over infrastructure budgets. The result is a road system that undermines the quality of urban life and crowds out other travel modes while new development areas are poorly served by transit and active transportation infrastructure such as sidewalks and bike lanes.

Moving away from automobile dependency requires greater coordination between land use and transportation planning at both the regional and local scales. Municipalities in Canada are responsible for transportation planning, but as we saw in chapter 5, they generally work within a policy framework laid down by provincial governments. Moreover, provincial governments provide much of the funding for major transportation projects and therefore have an important say over how a municipality's transportation priorities are realized in practice. Although provincial policy orientations vary widely across the country, it is probably fair to say that provincial governments use their influence over local governments to favor more transit-supportive development patterns to an extent greater than that found in the United States. Moreover, the less fragmented governance of Canadian urban regions has given rise to more integrated land use and transportation planning practices at the regional level than found in the United States, helping prevent suburban municipalities from taking a "free ride" on central cities and to maintain the financial health of regional transit systems.

Regional Transportation and Land Use Planning

Few Northern American metropolitan areas have been as consistent and successful as Portland, Oregon, in coordinating land use and transportation decisions. Metro, as the region's metropolitan government is known, has fought sprawl and automobile reliance with an integrated policy framework since the 1980s. It has extended public transit, revitalized older neighborhoods, and strengthened the downtown business district. Although there is much to learn from this experience, it is widely recognized Portland is an outlier in the US context; it is one of the only US jurisdictions with an effective metropolitan-level government. In contrast, Canada has experimented with a variety of metropolitan governance formats and mechanisms for coordinating transportation and land use planning at the regional level, providing a relatively rich store of experience from which to extract lessons for the United States.

Two urban regions in Canada with relatively successful transportation and land use coordination histories are Vancouver and Ottawa. In 1975, the Greater Vancouver Regional District (GVRD; now Metro Vancouver) adopted a strategic plan that called for concentrating growth in a series of regional town centers. These transit-oriented, mixed-use nodes would be linked together with a future rapid transit network (figure 6-1). The plan started to materialize in the 1980s as the SkyTrain network took shape, connecting several of the subcenters with an elevated, driverless train. The nodal concept was carried forward into the 1996 Liveable Region Strategic Plan, which called for the preservation

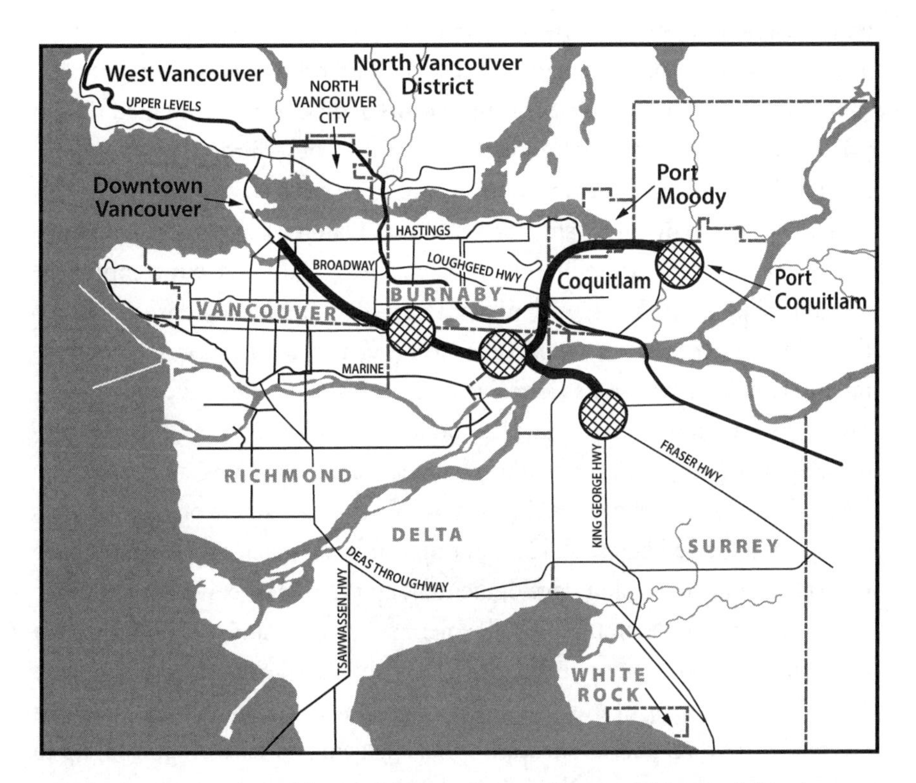

FIGURE 6-1. 1975 Livable Region Plan, Greater Vancouver Regional District
Source: Livable Region Plan 1975.

of a peripheral green zone (largely agricultural land) and the concentration of regional growth in and around the already urbanized areas of the region. As the land use plan was being prepared by the GVRD, the province launched a regional transportation planning process that called for the extension of the SkyTrain network to more subcenters. In recognition of the interactive nature of land use and transportation decisions, the two planning processes were pursued in tandem, and the resulting plans were mutually reinforcing.

This coordination between land use and transportation planning was continued into a new round of regional planning when the province created a regional transportation agency at the end of the 1990s and endowed it with regional transportation planning authority (see below). The agency—called TransLink—launched into a regional transportation planning exercise just as Metro Vancouver began to update its own strategic land use plan. The two resulting plans build on the nodal development strategy introduced in 1975 and add a new concept, the frequent transit network development area. These areas are within walking distance of SkyTrain corridors or high-capacity bus routes and are to be developed using transit- and pedestrian-oriented design principles. The new regional land use plan calls for two-thirds of new development to be located in either these areas or in town centers.

The Regional Municipality of Ottawa-Carleton (RMOC) was an upper-tier government created by the province in 1969 to carry out comprehensive planning, invest in regional infrastructure, and provide regional services to the sixteen (later reorganized into eleven) lower-tier urban, suburban, and rural municipalities within its extensive boundaries. The regional council oversaw transportation planning and appointed the board of the Ottawa-Carleton Regional Transit Commission, a unified transit-operating authority. The RMOC's 1974 official plan envisioned the Transitway, a network of rapid bus routes on dedicated lanes, and laid out the land use policies that would make it feasible: transit-oriented development along the routes, concentrated office development in a hierarchy of employment subcenters, and a requirement for regional shopping centers to be near the planned Transitway routes (figure 6-2).[2] The essentials of this vision have been preserved in subsequent plans created by RMOC and its successor, the amalgamated City of Ottawa, up to and including the current plan (adopted 2013).

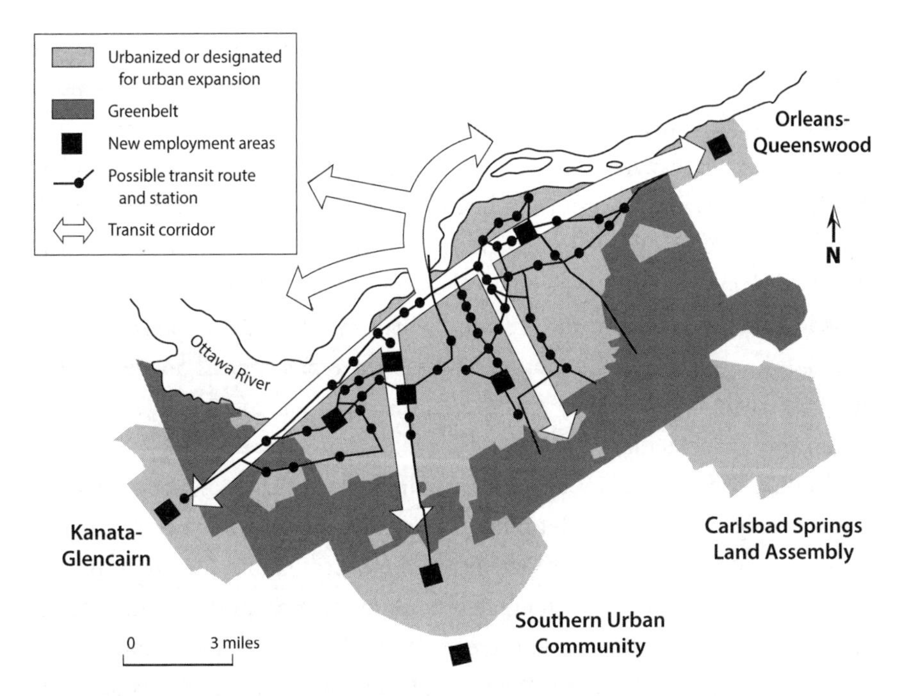

FIGURE 6-2. Urban structure plan in the Regional Municipality of Ottawa-Carleton Official Plan, 1974
Source: Adapted from Fullerton 2005.

Multimodal Transportation Planning

Some Canadian metro regions are equipped with multimodal transportation planning agencies mandated to work closely with regional land use planning authorities to manage demand for automobile use and shift demand to transit and active modes. Multimodal planning agencies have responsibility for planning and managing a region's transportation network across modes and municipal jurisdictions. They typically try to achieve a more balanced and integrated transportation system that improves mobility and accessibility through infrastructure improvement, demand management programs, and cooperation with land use planning agencies to promote smart growth policies.[3]

Vancouver region's TransLink, which has no counterpart in the United States, is a multimodal transportation agency created by the provincial government in 1998 to plan and operate transit services (including bus, the SkyTrain network, the Canada Line light rail transit system, and commuter ferry services), manage the region's major road network, and coordinate municipal efforts to expand the region's cycling network. It has an independent funding source (revenue from gasoline sold in the region and a portion of property taxes) and a mandate to expand the transit system in collaboration with Metro Vancouver and its regional land use plan. Facilitating this collaboration is that TransLink's board of directors, which manages the organization and prepares long-term plans, is appointed by the mayor's council, made up of all mayors in Metro Vancouver. This council also approves the agency's transportation and financial plans.

Toronto has been equipped with a multimodal regional transportation planning agency since 2006. Metrolinx was created by the province of Ontario with a mandate to provide leadership in the coordination, planning, financing, and development of an integrated, multimodal transportation network in the greater Toronto and Hamilton area. In 2009, the agency assumed direct responsibility for GO Transit, the regional commuter train and bus network, but municipal transit services are independently run. Unlike many similar organizations in the United States in which regional transit authorities operate in the absence of a strong regional land-use planning framework, Metrolinx is required to adopt plans and make investments in a way that conforms to the growth plan for the Greater Golden Horseshoe (see chapter 5). Its regional transportation plan, called the Big Move, was designed to reduce auto dependency and associated greenhouse gas emissions, boost transit and active transport modes, and integrate the region's many local transit systems. One of the plan's strategic goals is to contribute to the building of communities that are pedestrian-, cycling-, and transit-supportive, principally through a system of mobility hubs. These station areas, where transportation modes—including regional and local transit services, cycling and pedestrian networks, and car-sharing facilities—come together, are also locations for major destinations such as office buildings, hospitals, educational facilities, and government services.

In greater Montreal, the Province of Quebec formed the Metropolitan Transportation Agency (MTA) in 1995 to coordinate public transportation investment and services in the region. The agency has an independent source of income (from car registrations and gas taxes, plus a small share of property taxes collected by municipalities in the region). It works with the subregional transit providers (Montreal, Laval, and South Shore) to coordinate existing transit services and directly manages commuter rail and bus services linking Montreal to the suburbs. The agency is charged with regional transportation planning and has produced several plans since its founding. The current plan focuses directly on the need to link transportation planning with the management of growth in the region to achieve higher densities, mixed-use development, and a defined urban structure based on activity nodes. The Montreal Metropolitan Community (the regional level of government, which has land use planning powers; see chapter 5) has approval power over the MTA's transportation plan, which ensures a good fit between strategic land use and transportation planning in the metropolitan region. At present, the MTA is an agency of the provincial Ministry of Transport, although the province announced in 2015 that control will be transferred to a regional entity to be made up of mayors in the region and outside experts to be named by the government.

Transit-Oriented Site Planning

The disconnect between transportation and land use planning is obvious in the way that many cities in the United States plan for and provide transit services. Transit agencies usually have little or no influence over the land use patterns that determine demand for their services. Municipalities approve new low-density subdivisions or office parks without taking into account how residents or workers can gain convenient access to transit services, sometimes despite significant preexisting congestion. Moreover, transit provision to new communities, where it is provided, is usually delayed until sufficient build-out has been achieved to justify a new route. This delay allows incoming households to establish car-oriented behavior patterns and makes the shift to transit use more difficult. Low densities make all but peak-hour transit provision infeasible, yet studies have shown that peak-hour-only transit services are extremely inefficient and expensive.[4] Congestion is relieved by increasing system capacity for cars instead of transit. The divide between land use and transportation planning creates a situation in which sprawl and the associated car dependency emerge by default.

Although by no means spared these tendencies, the coordination of land use and transit planning appears to have been somewhat more routine in Canada than in the United States. This pattern has been noted by US observers for many decades. In his 1986 article on urban transit in Canada, Robert Cervero documented the careful level of integration between transit and land use planning that he believed, in addition to higher service levels, was behind Canada's higher transit use rates compared with the United States.[5] He was impressed by the integrated planning evident in the development of Metro Toronto's Yonge Street subway line in the 1950s and 1960s, a line that stretches from the central business district to the northern suburbs. Immediately following the completion of that subway line and in part due to the granting of density bonuses and issuance of air-rights leases, high-rise towers began mushrooming up around station areas. As a result, between 1952 and 1962, more than 90 percent of all office construction in Toronto (and half of all apartment additions) occurred within a five-minute walk of the Yonge Street corridor. In more recent decades, most office development has gravitated to car-dependent office parks in suburban locations, but the Yonge Street corridor (figure 6-3) continues to intensify and remains a vital hub for both living and working.

Just as land use regulation has been deployed to support transit services, transit investment has been used as a lever to guide urban growth in many Canadian suburban communities, often in advance of demand in support of regional plans. Extensions of both light and heavy rail lines in Toronto, Montreal, Vancouver, Edmonton, and Calgary have sought to cluster new office, commercial, and residential growth around designated satellite subcenters. In most of these cases, rights-of-way were reserved and protected far in advance of construction rather than waiting for development to take place and introducing public transit after station-area growth has already taken place. To maximize the effect on the location decisions of households and businesses, the future arrival of transit services is widely advertised. Meanwhile, zoning around stations is often adjusted to encourage higher-density development, and improvements are made to the physical infrastructure of the area so as to be able to accommodate more intense activity. These measures are in sharp

FIGURE 6-3. Intense development in Toronto's North York Centre, a mixed-use suburban node along the Yonge Street corridor, served by three subway stations *Source:* PFHLai, via Wikimedia Commons.

contrast to station areas in many US suburbs, where areas around stations are downzoned to preserve the low-density character of the suburban landscape.

Compared with US metro areas, Vancouver has very high ridership and transit modal share. In large part, its transit success has been due to the close integration between high-density mixed-use development in nodes around selected SkyTrain stations, from which the system draws a lot of its patronage. Park-and-ride facilities around stations in the metropolitan core cities of Vancouver, Burnaby, and New Westminster have been expressly excluded in favor of high-density uses clustered close to the station entrances. Transferable development rights and density bonuses have been used to funnel development to station areas. Due to high land costs, most parking in the town center districts is in multistory structures or underground, freeing up land for parks, passageways, and connecting bike paths. Municipal station-area plans and design guidelines have encouraged developers to pay special attention to the quality of urban design, giving rise to human-scale but dense precincts with good-quality bike and pedestrian access to transit facilities. For example, setbacks are typically limited along commercial streets near the stations, creating a more walkable environment. SkyTrain has had a significant effect on the development of areas near stations (figure 6-4); between 1991 and 2001, the population living within 500 yards of SkyTrain increased by 37 percent compared with the regional average growth of 24 percent.[6]

The provincial government has played a major role in the success of transit-oriented development around Metro Vancouver's SkyTrain stations. Its financial contributions to the SkyTrain system have been accompanied by strong direction to local municipalities to alter zoning and other codes to support compact, mixed-use development around stations. Relocation of government offices to town centers primed the development pump and attracted private investment. The province also assisted with the purchase, assembly, and servicing of lands around stations to ready them for redevelopment.[7]

Ottawa's Transitway system is an innovative rapid transit system based on dedicated bus lanes serving high-density employment subcenters throughout

FIGURE 6-4. Dense development around Metrotown SkyTrain Station, a suburban node in Vancouver
Source: freewindv7, flickr.

FIGURE 6-5. Employment development around a Tunney's Pasture Transitway Station in suburban Ottawa
Source: Steve Brandon, flickr.

the region. The system has more ridership per capita than any other similar-size transit system in Northern America. Many points of integration between land use and transportation planning have contributed greatly to the success of this system. At the site level, the transit authority, municipal planners, and developers have worked together to design new communities around transit stations rather than considering transit as an afterthought (figure 6-5). Transit rights-of-way are protected, and land is dedicated for future station use. City planners work with developers and the transit authority to shape development near the region's extensive transit lines, even in low-density suburban subdivisions. Following a set of provincial transit-supportive guidelines, land uses are distributed to take advantage of the proximity of transit services, with higher-density housing, office buildings, retail centers, and senior citizens' residences closest to the rapid bus stations. Collector streets are laid out to facilitate feeder buses, and sidewalks in residential subdivisions are required to facilitate pedestrian access to transit stops on main streets. Restaurants, banks, and day-care facilities are encouraged to locate in employment centers to minimize the incentive for employees to drive to consumer services. Park-and-ride facilities are limited to end-of-line stations so as to avoid compromising the pedestrian-friendly design around transit stations and to encourage riders to use feeder transit or walk or bike to the stations.[8]

Transit-Supportive Parking Policies

The availability and price of car parking is an important factor in the choice of transportation mode. As noted in chapter 4, the downtowns of Canadian cities tend to have significantly fewer parking spaces per employee than US cities. This difference reflects a deliberate policy aimed at reducing car use in the most congested central-city areas, where transit alternatives are readily available. Parking standards are the policy instrument of choice in this respect, with Canadian downtown areas having significantly lower minimum parking requirements for new development than in US cities.[9] Canadian cities also tend to have more restrictive parking policies around transit stations, whether downtown or not, and to use parking standards as a lever to encourage developers to incorporate transit-friendly provisions into real estate development projects along bus routes.

Calgary's downtown parking policy is a good example of how parking can work as a part of a wider strategy designed to discourage car commuting to the central business district and boost transit use. Since the 1980s, developers of all new buildings in the city center have been restricted to providing 50 percent of the bylaw-mandated parking on site, which for office buildings means about one-third of a parking space per 1,000 square feet of office space. Developers are required to make a cash contribution to the city's parking fund in lieu of meeting the remaining 50 percent of the parking minimum. The revenue from the fund has been used to build parking structures on the periphery of the city center, especially near rapid transit stations. As a final piece of the strategy, rapid transit within the perimeter of these parking structures was made free of charge. The overall effects of the strategy are to raise parking costs in the city center and encourage car commuters to switch to transit, at least for the last leg of their journey to work.

Many other cities in Canada provide exemptions or relaxations of the minimum parking requirements for buildings within transit-rich downtowns or a certain distance of a transit station. In Ottawa, parking requirements were eliminated for higher density residential development in the downtown core, and other types of development were permitted to meet part of their parking requirements through shared parking (e.g., a restaurant that needs parking in the evening and an office building that needs parking during the day). Instead of minimum parking, Toronto's zoning ordinances set a maximum number of parking spots per unit of office or retail floor space. In St. John's, Newfoundland, parking requirements in the majority of the downtown area were eliminated for most land uses in the 1980s.

Parking strategies have also served to help build ridership for transit systems in Canadian cities by limiting parking near transit stations. The predominant land use around many suburban rail stations in the United States is parking lots. In Toronto, in contrast, mid- or high-rise housing huddle around most stops, and some are flanked by commercial centers. Toronto has restricted parking mainly to terminal stations, which serve a potentially large catchment of suburban and exurban residents commuting to the city. As a result, Toronto averages far higher shares of walk-and-ride, bus-and-ride, and bike-and-ride customers at its suburban stations than found in US rail cities. Parking supply is also capped for development projects near rail stops, especially in major subcenters. For example, around the North York Centre subway station, located in a suburban center within the City of Toronto, parking is limited to one-third of a parking space per 1,000 square feet of office space (as in Calgary), a fraction of the office space required in many suburban office complexes in the United States.[10]

When Ottawa's Transitway opened in 1983, the federal government—the region's biggest employer—began eliminating free parking for its employees and reducing parking availability in its buildings near the rapid bus routes. Meanwhile, the city lowered minimum parking standards for employment and residential uses within 650 yards (walking distance) of a Transitway station. For many years, the regional transit authority has restricted park-and-ride facilities to the Transitway's terminal stations to encourage the use of feeder and express buses. In Calgary, parking requirements were reduced for multifamily residential developments located within 650 yards of a rapid transit station or within 500 feet of frequent bus service. In Edmonton, the city will relax minimum parking requirements for any project near transit if the developer can produce a parking demand study that shows that fewer parking spaces are needed.

Another strategy employed by Canadian municipalities is to use parking requirements as a lever to encourage developers to build more transit-friendly projects. The City of Ottawa, for example, has allowed a reduction of twenty-five parking stalls for every bus stop provided at retail centers. Even some suburban municipalities have adopted innovative parking strategies to enhance transit use. Since 1997, Richmond Hill, a suburb north of Toronto, has offered reduced parking standards in new developments in exchange for transit-friendly amenities. The zoning bylaw stipulates that developments on transit routes can see their parking requirements reduced if transit facilities are incorporated into the development. Easy pedestrian access to bus stops and bus shelters can reduce parking requirements by as much as twenty-five parking spaces. Grocery stores located on arterial roads are considered ideal candidates for this program. For major development proposals, such as a regional mall, the municipality may require on-site transit terminals in exchange for parking credits.

Transit Service Levels

Better integration between land use and transportation planning can help explain higher transit use in Canada than in the United States. Many of Canada's urban regions have been planned in a way that would facilitate transit, with a strong city center and subcenters positioned on high-quality transit routes. Even in neighborhoods remote from such transit services, community design has tended to favor transit-supportive features, with higher densities, less surface parking, and better pedestrian facilities.[11] As Paul Mees has argued in *Transport for Suburbia*, however, differences in urban form alone cannot explain differences in transportation outcomes between cities.[12] Although he admits that density plays a role in the success of transit systems, he points out that cities with high densities do not necessarily have good transit systems. For example, Ottawa's density is a third lower than that of Los Angeles, but transit use is four times higher (and, incidentally, walking is three times as high and cycling four times as high). Rather than urban form, he attributes the different levels of ridership between cities to differences in transit service levels.

Toronto and Chicago provide a good contrast to illustrate this point. The Toronto Transit Commission (TTC) and the Chicago Transit Authority (CTA) were in similar circumstances in the mid-1950s, with a quality transit system serving a high-density core and a series of inefficient private bus or train lines providing minimal services to the surrounding suburbs. When Metro Toronto was formed in 1954, the TTC did not use low suburban densities as an excuse for doing nothing (as the CTA did), but instead bought out the suburban bus companies and installed a system of high-frequency bus routes feeding riders into the newly opened subway line. As the subway system expanded and suburban development continued, new feeder lines captured more riders and reversed the decline of ridership in the 1960s, despite the growth in incomes and levels of car ownership. The rapid transit spine continued to expand until 1985, and ridership grew in step until a serious recession in 1990 forced the TTC to cut service levels, especially to the bus routes, and raise fares. The system entered a spiral of decline in the 1990s, exacerbated by the elimination of provincial subsidies. In the first decade of the twenty-first century, lost patronage was partially regained as services improved with the advent of modest federal funding (from a portion of the federal gas tax) and resumption of provincial assistance.[13]

A similar story can be told for other Canadian cities. In Ottawa, transit use in suburban areas is high compared with US suburban areas, which can be attributed to generously supplied and carefully configured bus service integrated across the whole urban region.[14] In particular, service frequencies on city streets are high, and routes are relatively closely spaced and well-integrated, with the Transitway buses running on dedicated lanes. Thus, almost everyone in Ottawa, including those who live in relatively low-density subdivisions, has access to good-quality transit service. Similar transit supply strategies are seen in Vancouver, Calgary, and Montreal.

Zooming out to the national picture, we find further evidence that variations in transit services help explain changes in ridership levels. Figure 6-6 tracks both public transit service and ridership levels for the two countries for the post-World War II period. It shows that transit service declined steeply in both the United States and Canada in the 1950s. Canadian service started to increase in 1966 and then shot up in the 1970s, after which it stabilized during the 1980s. It was an era of massive investment in high-capacity new technology in cities throughout Canada, including Toronto's and Montreal's world-class subway systems, Calgary's and Edmonton's pioneering efforts to build light rail system, Vancouver's advanced SkyTrain, and Ottawa's rapid bus Transitway. In the 1990s, service levels declined in Canada as the population grew without major new investment, but this trend turned around at the end of the 1990s with renewed investment from both provincial and federal sources. Since then, there has been a significant increase in service levels.

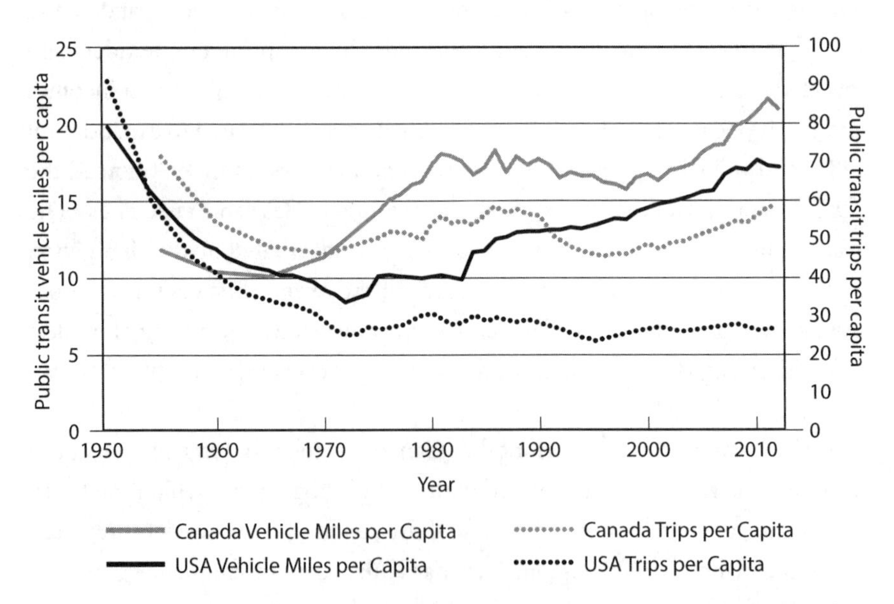

FIGURE 6-6. Annual public transit vehicle miles of service and public transit trips per capita, United States and Canada, 1950–2012
Source: American Public Transit Association, Public Transportation Fact Book, 2014.

Meanwhile, in the United States, transit service levels declined until the early 1970s and stayed low into the 1980s, creating a major gap with the climbing levels in Canada as shown in figure 6-6. The greatest difference between the two countries was in 1982, when Canadian service levels were almost 80 percent higher than those in the United States. From then on, service levels in the United States began to rise steeply, reducing the gap between the two countries, but by 2012, Canadian service levels were still 22 percent higher than in the United States.

Comparing these trends to the trends in ridership reveals an interesting difference between the two countries. Although transit ridership seems to respond directly to changes in service levels in Canada, the response in the United States is sluggish. In Canada, ridership began a slow decline at the end of the 1980s as service levels began to decline and then picked up in sync with service levels when they began to rise in the later 1990s. In contrast, despite almost doubling per capita service levels in the United States from the late 1970s to the present, ridership has been stubbornly low; as seen in figure 6-6, after a gradual decline until the mid-1990s, per capita patronage of US transit systems has recovered to about the level it was in 1971.

What explains the differential outcome of transit investment between the two countries? To some extent, it reflects differences in spending strategies over the years, with Canada putting its resources into inexpensive bus service improvements and light rail while the United States spent more on very expensive commuter rail that had disappointing results in terms of ridership gains. Transit demand modeling has shown that factors external to the transit system itself are also important in explaining why smaller increases in service levels in Canada will result in greater ridership gains than in the United States.[15] Foremost among these external conditions is the pattern of urban development. More-compact Canadian development patterns entail more transit services on offer within a smaller geographical footprint, making transit service more attractive and efficient. Running new buses or light rail into areas that can support transit is a more effective strategy than building heavy rail commuter lines—as has often been done in the United States—to distant suburbs that are already built around the car.[16]

With this conclusion, we have come full circle. Density cannot alone explain the differences in ridership between the United States and Canada, but it does play an important role in combination with how transit dollars are spent. Better coordination between land use and transportation planning in Canada over the decades has given rise to an urban fabric that can support transit, and when transit investments are made, transit ridership rises.

Highway and Transit Financing

As we saw in chapter 4, the United States has substantially more urban freeways on a per capita basis than does Canada. The evidence shows that highway investments in urban areas can contribute to a sprawling urban form and lead to higher levels of energy use. Highways also contribute to the decline of central cities, which has equity, livability, and sustainability consequences. Major investments in highways lock metro regions further into the cycle of car dependency and development patterns designed around private rather than public or active transportation.

Given the serious implications of highway development for urban sustainability, it is important to understand how the different levels of highway provision have arisen between the two countries. The level of car ownership has undoubtedly played a role here; as we saw in chapter 4, the United States has consistently had higher levels of car ownership in the post–World War II years than Canada. Car owners create a natural constituency for highways, and elected officials are aware of the political points to be scored by announcing new highway infrastructure that will (at least temporarily) address congestion in a given corridor. More highways encourage more people to purchase cars and migrate to car-dependent locations, creating a growing basis of support for the highway lobby to demand additions to the freeway system in a self-reinforcing cycle. Higher incomes in the United States and lower gas prices than Canada might also play a role in explaining higher levels of US car ownership and use and therefore the greater interest in highway development.

Although these factors are undoubtedly important, most observers of Canada-US differences have tended to point to another, less obvious factor: the way in which highways are planned and financed in the two countries. In the United States, the system of limited-access highways took root in the 1930s, but not until the 1950s, when the federal government undertook its vast program of highway building under the Federal Aid Highway and Highway Revenue Acts of 1956, did it begin to flower. Under the initial plan, a vast system including not only intercity trunks but also urban radials and circumferential arteries designed to enhance intrametropolitan access was authorized. By 2006, the interstate system amounted to almost 47,000 miles of roadway, the largest system of highways in the world and probably the largest public works project in history.[17] The cost of constructing the network was estimated at $425 billion in 2006,[18] which would be about half a trillion dollars today.

The financing formula behind highway investment has been criticized for contributing to the oversupply of highways in the United States. First, federal contributions come from the Highway Trust Fund, which is based largely on the federal portion of the taxes levied on gasoline and diesel fuel throughout the country. The fund provides a dedicated source of revenue that has been overwhelmingly used for expanding and maintaining the highway system in a self-reinforcing cycle, with few competing priorities. Most states have followed suit, with laws requiring that revenues from the state portion of vehicle fuel taxes be used largely or exclusively for highway construction, maintenance, and operations. Second, federal aid has covered as much as 90 percent of the cost of building freeways, with the remaining 10 percent covered by the state. Critics claim that this approach has skewed transportation investment in favor of freeway building because few state and local authorities are likely to pass on the opportunity for significant federal aid when so little additional funding is required.[19]

Another concern is that the central role of the federal government in paying for and building highways has spurred urban highway construction without proper consideration for local consequences, including changes in urban form and public transit demand. The same distortion has been observed with federal transit funding, which emphasizes capital projects and leads to the overproliferation of expensive urban rail projects. Although the federal government does require projects to demonstrate threshold projected ridership levels as a condition of funding, these investments have often returned disappointing performance results.[20] In most urban areas, metropolitan planning organizations (MPOs) coordinate federal funding for highways, transit, and other transportation infrastructure and are sometimes part of regional councils of governments (COGs) that prepare long-range land use plans. These arrangements, however, have proven ineffectual as a mechanism for coordinating transportation and land use at the regional level (see chapter 5).[21]

In 1991, the United States began to consolidate highway and transit funding. The Intermodal Surface Transportation Efficiency Act (ISTEA) permitted states to divert highway funding to transit projects. ISTEA was succeeded by TEA-21 in 1998 and SAFETEA-LU in 2004, each of which retained the overall flexible approach. The extent to which highway funding was redirected to transit after ISTEA passed varied widely by state, but overall funding was still heavily skewed toward highways. In a few cases—including Salt Lake City, Denver, Dallas, Charlotte, Las Vegas, San Jose, and San Diego—MPOs used the ISTEA/TEA-212 process to increase the funding of light rail systems, but those are the exceptions that prove the rule. In most metropolitan areas, weak and understaffed COGs and MPOs have not been able to fight state highway departments and the many advocates of continued highway construction.[22]

Canada has a different process for providing urban transportation infrastructure, one that is frequently cited as more neutral with respect to transit and highways. The only highway project on a national scale in that country has been the Trans-Canada Highway, for the most part just a two-lane surface road stretching from Victoria, British Columbia, to St. John's, Newfoundland. The terms of federal participation in this project were legislated in the 1949 Trans-Canada Highway Act, which committed the federal government to sharing at least 50 percent of construction costs with the provinces for seven years to a maximum of $150 million. There were multiple amendments to the act as the scope of the commitment grew, and by 1971, when the project was finally completed, the federal portion had amounted to $825 million (in nominal dollars).[23] The amount represents about 0.2 percent of the federal contribution to the interstate system in the United States.

Aside from the Trans-Canada Highway, the provinces and municipalities share most of the responsibility for planning and funding urban transportation infrastructure in Canada, with only sporadic ad hoc contributions from the federal government. Gasoline taxes in Canada at both the federal and provincial levels historically have been paid into general revenue funds, not tied to highway or road funding.[24] This arrangement compels provincial and local governments to weigh the benefits and costs of highway investments against competing funding priorities. It may also help explain why Canadian cities have fewer urban freeways.[25]

The absence of a nationally funded urban highway program in Canada has also led to more integrated land use and transportation planning at the regional level.[26] Given their role as cost-sharing partners on capital projects and their constitutionally enshrined involvement in urban affairs, Canadian provinces can use funding decisions to leverage supportive land use planning practices at the local level. In addition, because they have more of a stake in the game, local and regional governments are motivated to coordinate infrastructure investment decisions with supportive land use policies, such as upzoning land around planned transit stations. Finally, in contrast to the anemic metropolitan planning organizations found in most US metropolitan areas, many of Canada's metro areas have effective planning practices that help integrate land use with transportation investment on a regional scale.

The flexible funding of transportation infrastructure in Canada has contributed to major highway project cancellations and a shift of funding resources to transit. In Toronto, the Spadina Expressway would have cut a swath through inner-city neighborhoods. Opposition to the freeway, led in part by American urbanist Jane Jacobs, who had moved there from New York in 1968, caused Ontario Premier Bill Davis to cancel the project just before the 1971 election and funnel the funds into public transport instead. The increased funding was used by the TTC to push subway lines into suburban areas where densities had previously been regarded as too low to justify rail transport. The frequent suburban bus service lines inaugurated in the 1960s helped overcome the density deficit and ridership in these areas boomed.[27]

In Vancouver, a plan for an extensive freeway system was scrapped when antifreeway activists won control of the city council in 1972, one reason Vancouver has almost no highways within the city limits. The newly formed regional planning agency, the GVRD, began work on a regional transport and land use plan as an alternative to the canceled freeway system. Inspired by the success of the Metro Toronto model, the 1975 GVRD plan relied on a regional light rail system serving the entire urban area, not just densely populated inner regions, and region-wide bus services to feed riders from suburban areas into the light rail transit system. A regional public transport bureau created by the province took control of City of Vancouver buses and trolley-buses along with private and municipal routes in the suburbs. The authority integrated and expanded regional bus services and began to plan a light rail system. Public transport usage rebounded, with substantial gains in mode share in suburban areas. The first light rail line opened in 1986, and the system has grown steadily ever since.

Vehicle Ownership and Operation Costs

Another factor that contributes to higher transit and biking rates in Canada is the greater cost of owning and operating a motorized vehicle there compared with in the United States. Sales taxes on new and used cars, shipping and import fees, vehicle registration costs, and licensing fees are generally higher in Canada than the United States. For instance, in the province of Quebec, federal and provincial sales taxes amount to 15 percent on the cost of a new car, or about $5,000 on an average sedan. Of paramount importance in this regard, however, is the price of gasoline, which has been shown to shift the economic calculus toward transit and bikes where these modes present feasible alternatives to the private car.[28]

Low-density suburban landscapes are intense energy users. Many critics of sprawl and car dependence note that the extreme suburbanization around US cities would not have been possible without abundant and low-cost energy during the post-World War II period.[29] Low gasoline prices encourage automobile ownership, less efficient vehicle fleets, and longer average distances driven. Energy prices in the United States have been historically much lower than those found in other industrialized countries where growth has been more compact, including Canada.[30]

Figure 6-7 shows the retail price of gasoline in the United States and Canada since 1950. Prices were relatively stable until the early 1970s, when they rose dramatically due to international events that restricted the supply of oil to the West. In the 1980s, when the supply of oil on the world market increased, oil prices fell. At this point, pump prices in the two countries began to diverge

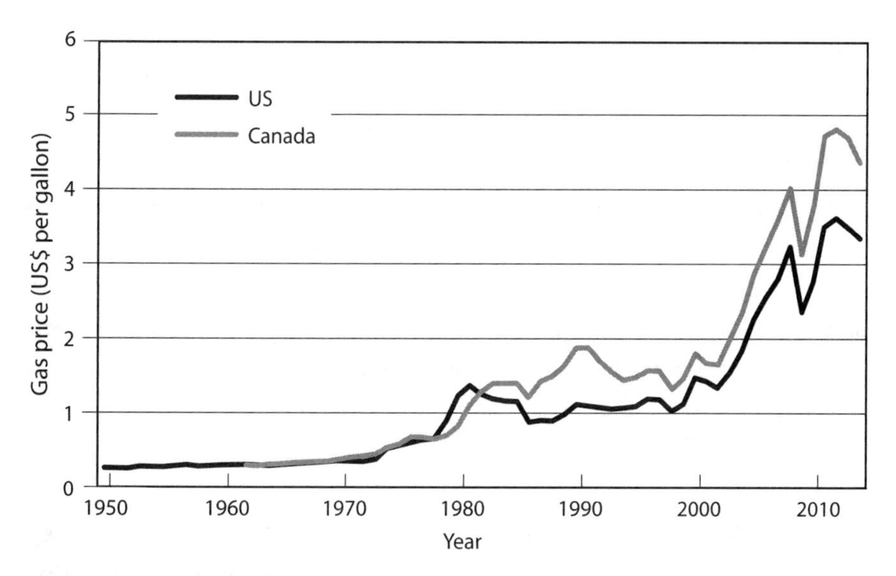

FIGURE 6-7. Consumer price of gasoline, 1950–2014
Sources: US Energy Information Administration—Retail Motor Gasoline Annual Average Prices, 1949–2001l; Natural Resources Canada, Average Retail Prices in Canada 1987–2015; other data courtesy of Paul Schimek.

as Canadian governments (federal and provincial) imposed heavy consumer taxes on gas sales, taking advantage of falling prices to smooth the way for tax increases. In the United States, gasoline taxes were also increased during this period, but not by nearly enough to make up for the decline in market prices. Since then, the consumer price for gasoline in Canada has remained well above that in the United States, with gasoline taxes that are twice as high in Canada as in the United States.

The persistently higher cost of gasoline in Canada has likely affected the development and travel patterns found in its cities. Motorists can compensate for short-term rises in gas prices by curtailing discretionary driving or purchasing vehicles with higher fuel efficiency. Sustained higher prices, however, lead to more fundamental shifts in location choices and travel behavior; people tend to move closer to workplaces to reduce the length of their commute and switch to alternative modes of travel such as biking, walking, and public transit.[31] All these patterns are evident in Canada.

Conclusion

Canada and the United States have many things in common that account for the high levels of automobile dependence in their cities compared with those of Western Europe. In the years after World War II, the two countries enjoyed high levels of prosperity, witnessed the mass introduction of the automobile, experienced dynamic population growth, and had ample land on which to spread out. Despite these similarities, as we saw in chapter 4, Canadians average about twice as many transit rides per capita as their US counterparts, own fewer cars and drive them significantly less, and bicycle at a rate three times that found in US cities. As a result, Canadian cities produce fewer greenhouse gases and use less energy for transportation than US cities.

The issues we have explored in this and the previous chapters help account for these differences and perhaps provide some insight into policies and approaches that may be applicable in certain contexts in the United States. In chapter 5, we showed that although the federal government in Canada has little influence on urban and transportation planning, the provinces exert considerable control, using their power to reorganize municipalities and create metropolitan governance regimes. In this chapter, we have seen how regional governance structures are able to effectively integrate land use and transportation planning across entire urban regions. This regional approach is in contrast with practices in the United States, where most state governments have ceded land use and transportation planning controls to local government and where the federal government has played an inordinate role in supporting highway development.

The ability to guide land use and plan transportation at the metropolitan level has allowed Canadian cities to foster more compact growth and, in at least some cases, focus growth in areas well served by transit. These development patterns have improved the economic viability of public transport and enhanced the effect of transit investments on ridership. Unlike in many regions in the United States where regional transit authorities operate in the absence of a strong regional land use planning framework, multimodal transit agencies in Canada tend to work within a regional planning framework that allows greater coordination between land use and transportation planning. Transit agencies, working with regional and local planners, have been able to proactively prepare the way for extensions of rapid transit systems so as to maximize the effect on growth potential in station areas and along transit corridors.

Urban planning practices in Canada have not prevented or even opposed suburbanization, but the type of suburb that is built can be quite different than that found in the United States. We have seen that transit service levels are much higher in Canadian metropolitan regions, with local bus services carefully configured to support rapid transit, even in suburban locations. Except in the most car-dependent subdivisions, bus service frequencies are relatively high and routes are spaced such that most residents have access to transit services. Growth energies are partially channeled into mixed-use centers that are well supported by transit services. At the site level, stronger direction provided by provinces and regional planning agencies has helped align local policies to support transit in suburban areas, with appropriate urban design and zoning for higher densities, a mix of land uses, and more restrictive parking policies around transit stations.

In central cities, the relative dearth of highways and more restrictive parking standards have favored transit commuting to downtowns while also helping maintain city centers as vibrant, economically viable, and pleasant places to live. Unlike so many US cities, Canadian central cities have not been abandoned by middle-class households, which has important consequences for transport. As in Europe, public transport serves a broad socioeconomic spectrum of the Canadian population; thus, it can draw on a large potential market and count on the political support of a wide swath of social strata.

Many of the policies and practices that have built support for transit in Canadian cities have also helped support bicycle use and walking. The denser, mixed-use development patterns in Canadian cities lead to average trip distances that are only half as long and thus more bikeable than the longer trips made in US cities. Canadian cities have done more than US cities to facilitate cycling by providing more extensive networks of separated bike paths and copious amounts of bike parking. These measures improve cycling safety in Canada relative to the United States, which in turn encourages more cycling there. The higher cost of owning and operating a car in Canada compared with in the United States also helps shift demand to active transportation modes for shorter trips.

The governance structures, urban and transportation planning policies, urban form, and transportation system variables that we have explored in this chapter all contribute individually to a more balanced transportation pattern in Canadian cities. In combination, though, they appear to reinforce one another and amplify their effects. Needless to say, this "virtuous circle" of self-reinforcing smart growth features does not operate ubiquitously throughout Canada. Some urban regions in Canada have had less success than others in their struggled against the decentralizing forces of sprawl, and even the more successful regions are still marked by landscapes of low-density, auto-oriented growth that would look somewhat familiar to US visitors.

Some observers have argued that because Canadian cities still have sprawl and associated problems similar to those found in the United States, the policy differences between the two countries make little difference.[32] Sprawl, these observers conclude, is inevitable given the economic, technological, and cultural forces at work, and no policy choice will alter this geographical destiny. This argument misses the main point, however: its not that there is no sprawl in Canada, but there is manifestly less of it than in the United States. Governments in Canada have managed to temper the forces of sprawl and make their cities more sustainable than those in the United States. The value of the Canadian experience is that it suggests that public policies complementary to transit, biking, and walking can be implemented and proven effective, even under economic, geographical, and demographic conditions similar to those found in the United States.

Chapter 7

Social Policy and a More Inclusive Society

Although social policy is often seen as a subject independent of and largely unrelated to land use and regional growth, the two are in fact closely related. Social policies have had a strong influence on the course of both urban and suburban development in both the United States and Canada, although often in ways that are not immediately apparent. That is particularly true when one looks at the ways in which the trajectories of US and Canadian central cities have diverged since the end of World War II.

In this respect, two key social policy values tend to be given substantially greater weight in Canada; the first we call equalization, and the second is inclusion. In particular, as we will see, Canadian social as well as fiscal policy is strongly oriented toward equalizing differences, whether in fiscal resources, educational opportunity, or otherwise, that in the United States are often allowed to remain unchecked (although with considerable variation from state to state). For that reason, the discussion of Canadian policies in this chapter begins with a discussion of the national fiscal equalization policy. Although that policy may seem to be rather far removed from the realm of social policy, we believe that the connection will become clear. From equalization, we then look at social support, immigration, housing, and public education.

Equalization

In chapter 2, we quoted a commentator who said that "things like equalization and healthcare bind a lot of the country together."[1] What does that mean, and why is it such an important part of the Canadian system? Equalization is a central element in the Canadian federal system, a program by which federal revenues are distributed to the provinces on the basis of a formula that measures the disparity between each province's ability to raise revenues and equalizes the fiscal resources available to each province to provide services. Under equalization, the federal government distributed just over $16 billion[2] to six "have-not" provinces in the 2013–2014 fiscal year, nearly half of which went to Quebec.[3] This amount is roughly equivalent to the US government distributing nearly $150 billion to the nation's less affluent states.

The rationale for equalization payments is that they "enable less prosperous provincial governments to provide their residents with public services that are reasonably comparable to those in other provinces, at reasonably comparable levels of taxation."[4] In other words, although the payments are unconditional and no attempt is made to define such a fuzzy concept as "reasonably comparable," the underlying policy intent is clear; it is, as the same commentator pointed out, to ensure that "that if you live in Newfoundland you should have roughly the same access to healthcare and quality of life as someone living in, say, downtown Calgary."[5]

Equalization in Canada is intrinsic to the fundamental workings of government. In one form or another, it has been in existence since the creation of the Dominion of Canada in 1867 and was enshrined formally in the Canadian Constitution of 1982. That is not to suggest that there are no differences over the manner in which equalization is handled and who gets what; on the contrary, they are matters of periodic dispute and conflict. It is fair to say, however, that the principle of equalization is sacrosanct.

The principle of equalization, however, goes beyond the core federal program. The federal government provides additional transfer payments in the form of the Health Transfer program, through which $30.3 billion was transferred to the provinces for health care costs, and the Social Transfer program, under which an additional $12.2 billion was allocated to the provinces specifically to help cover social services, post-secondary education, and early childhood education and care. As figure 7-1 shows, the level of federal transfers for health and social services over and above the core equalization program has more than doubled since 1993 and is continuing to grow. Including a number of other, much smaller programs, the federal government disbursed $62.3

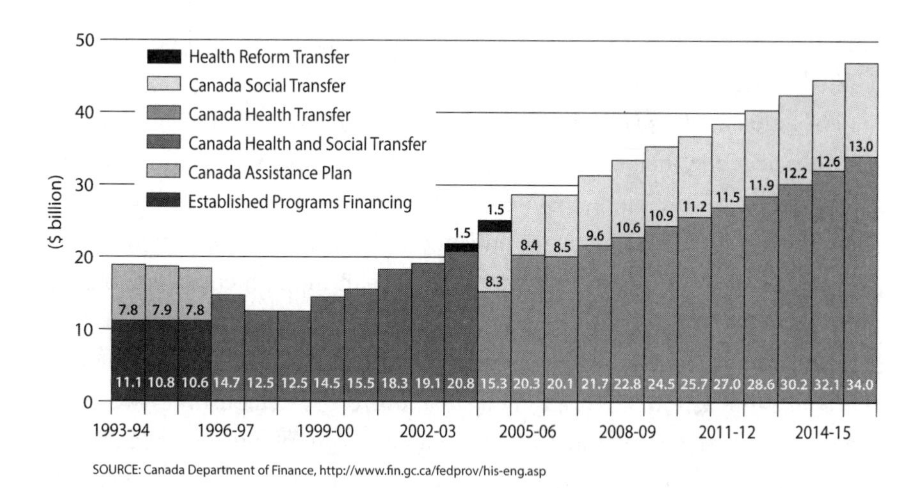

FIGURE 7-1. Health and social transfers from federal government to provinces, 1993–2016
Source: Canada Department of Finance, History of Health and Social Transfers.

billion to the provinces and territories in 2013–2014, equivalent to what would be nearly $600 billion in the United States.

The United States briefly implemented a revenue-sharing program not unlike equalization from 1972 until it was abolished under the Reagan administration in 1986, although it was on a far more modest scale, generally in the area of US$5 billion per year. Today, many programs of the US government have an equalizing effect, and state-by-state comparisons between federal outlays and federal tax revenues received show wide variations. Generally speaking, these variations favor poorer states if only because the progressive nature of the federal income tax tends to mean that states with wealthier taxpayers shoulder a higher share of the federal expenditure burden. The significant differences between US and Canadian policy, however, are many. First, unlike Canada, the United States has no explicit policy of equalization. Second, whatever equalization takes place in the United States, it is neither transparent nor coherent. Third, the extent to which equalization in the United States is a function of political log-rolling and the efforts of powerful politicians to seek their share of military spending and other benefits for their states. Thus, although equalization plays a major explicit role in

Canadian policy, it plays no comparable role in fiscal or social policy in the United States.

The concept of equalization is similarly central to the relationship between the provinces and their constituent municipalities. The fiscal relationships between provinces and municipalities in Canada have also had the effect, and in many cases the explicit intent, to enhance social equity in metropolitan areas and reduce local variations in poverty and wealth through equalization of resources, including reorganization of governmental structures within metropolitan areas so as to reduce fiscal inequalities among individual municipalities. Through annexation, through amalgamation, or by establishing two-tiered government systems, provinces have combined central cities with their suburbs to ensure that the benefits and burdens of urban growth and change have been more equally shared across the metropolitan area. Such reorganizations have allowed metro areas to more fairly apportion investment in social housing so as to prevent overwhelming concentrations in specific neighborhoods and to share the costs of municipal services and metropolitan installations more equitably between central cities and suburban districts.[6]

Provinces also transfer much of their support to municipalities through unconditional grants and award them according to formulas that took into account variations in the ability of different municipalities to meet the cost of providing services. Ontario's 2015 provincial budget provides $515 million in its Municipal Partnership Fund for this purpose.[7] Funds are allocated on the basis of revenue-raising capacity, with adjustments for "challenging fiscal circumstances." Saskatchewan dedicates one point of the provincial sales tax to municipal revenue sharing, which amounted to $257 million in the 2015 provincial budget, of which $85 million went to the province's two largest cities, Saskatoon and Regina.[8]

Equalization payments are coupled with provincial governments assuming increasing responsibility for services previously largely delegated to municipal governments, including the administration of justice, education, health, and social services, a practice known as provincial uploads. The Ontario 2015 budget identifies savings to municipalities of $1.7 billion in cumulative benefits of provincial uploads enacted since 2008 in areas such as disability support, drug benefits, and short-term assistance to people in financial need (Ontario Works). In addition to the Partnership Fund and the provincial uploads, the provincial budget provides an additional $1.6 billion in municipal assistance, principally for health-related costs. The picture in this respect is not quite as positive as these figures suggest, however, because at least some of the upload benefits appear to be compensating for the effect of extensive downloading of costs from Ontario and other provinces to the municipalities that took place during the 1990s, which in turn reversed in part a trend toward uploading services evident in the 1960s and 1970s. Still, Canadian provinces generally play a much stronger role in the provision of social services than do most states in the United States, and their levels of social investment are generally higher.

Frances Frisken, a distinguished Canadian urban studies scholar who has written extensively on this subject, concludes that taken as a whole, provincial policies have evinced "a greater commitment on the part of provincial as compared to [United States] state governments to recognize and overcome disparities in the fiscal capacity, service requirements and servicing capabilities of municipalities that have fared unequally from ongoing processes of economic development and decline."[9] Fiscal equalization is not unknown in the United States. A few states have tackled this issue, particularly with respect to the costs of public education, but usually only after being ordered to do so by state courts.[10] The overall record is a bleak one, however; fiscal disparities have fostered the well-known "race for the bottom" dynamic, pitting municipalities against one another as they compete to attract development and further exacerbating differences in fiscal capacity and service equality across municipalities within metropolitan areas.

The Social Safety Net

Although there has arguably been some erosion in recent years as the country has been affected both by the pressures of globalization and a succession of right-wing national governments, Canada's social safety net still remains more robust than that of the United States. The existence of a single-payer health care system, in which all Canadians are guaranteed largely free health care, has been widely recognized as both more effective as a means of delivering health care and more efficient than the complex, and all but incomprehensible except to experts, US system. It is also popular. A 2003 poll found that 57 percent of Canadians were satisfied with the availability of affordable health care compared with only 25 percent of Americans,[11] and a 2009 poll found that 70 percent of Canadians believed that their system was performing well and that 82 percent preferred it to the US system.[12] In important respects, the Canadian system appears to provide better health outcomes at lower cost than that of the United States, as shown in table 7-1.[13] Although Canada spends 11 percent of GDP on health care compared to 18 percent in the United States, life expectancy is longer, and both infant and maternal mortality rates are lower in Canada.

The systems that provide support to families and individuals in need are also different. Canada provides a substantially more extensive system of both child benefits and old age benefits than the United States, including Old Age Security and Guaranteed Income Supplement programs for the elderly in addition to the Canada Pension Plan, which is analogous to the Social Security system in the United States.

Although precise comparisons are difficult, the available information indicates that a much smaller percentage of the Canadian population, as well as a much smaller share of Canadian children, live in poverty compared with their United States counterparts. Canada does not have an official definition of poverty; instead, a number of different measures exist, of which the most widely used is known as the Low Income Cut-Off after Taxes (LICO-AT).[14] Using this measure, which appears to cut off at a higher level than the official US poverty line,[15] 8.8 percent of Canadians were below the line in 2011, ranging

TABLE 7-1.
Comparative health indicators for Canada and the United States

	Canada	United States
Life expectancy at birth	81	78
Maternal mortality rate (per 100,000 births)	6.5	12.7
Infant mortality rate (per 1,000 births)	5.1	6.8
Per capita health expenditures (US$)	$4676	$8,467
Health expenditure as a percentage of GDP	10.9%	17.7%
Government health expenditures as a percentage of total government spending	17.4%	20.3%

Source: World Health Organization.

from 10.7 percent in British Columbia to 4.4 percent in Prince Edward Island; in 2011, the poverty rate in the United States was 15.9 percent.[16] Notably, less than 20 percent of single-parent families were below the LICO-AT in Canada compared with 37 percent of single-parent families in the United States falling below the poverty line. By another method of calculation, the percentage of households earning below 50 percent of the median income, 12.6 percent of Canadian households fell under that line compared with 25 percent of households in the United States.

Further reduction in the number of Canadians living in poverty has become an important policy goal; as of 2013, all but two of Canada's provinces and territories (British Columbia and Saskatchewan) had adopted official poverty reduction strategies, beginning with Quebec in 2002 and Newfoundland and Labrador in 2006.[17] Many of these strategies were grounded in explicit targets, such as the goal of Newfoundland and Labrador to "transform the province into the one with the lowest rate of poverty in Canada by 2014" or that of Ontario "to reduce the number of kids living in poverty by 25 per cent over the next 5 years."[18] Although the results of these strategies to date appear mixed, the nature of this initiative reflects the existence of Canada's continuing commitment to social inclusion.

Finally, Canada has a substantially higher minimum wage than the United States. Although there is no federal minimum wage in Canada, the provincial minimums range from $10 per hour in New Brunswick and the Northwest Territories to $11 per hour in Ontario and Nunavut. That, in turn, may also be affected by the much higher level of union membership in Canada, as noted in chapter 2.

Immigration Policy

As we mentioned in chapter 2, immigration policy is another important area in which Canadian policies have been different from those of the United States, with significant implications for the central cities of the two countries. Although both countries are important immigrant destinations, Canada has steadily absorbed a significantly larger flow of immigrants relative to its total population since the 1950s, something that may surprise Americans who are accustomed to the acrimonious debate about the number of immigrants being admitted to the United States. Foreign-born residents make up 21 percent of Canada's population compared with 13 percent of that of the United States. Figure 7-2 shows the average annual number of immigrants admitted to permanent resident status in the United States and Canada since World War II *as a percentage of each country's base population*. As the figure shows, even during Canada's low point in immigration during the 1980s, its relative share of immigration was significantly higher than the United States.

Notable, however, is that during the years in which cities in both the United States and Canada were under the greatest stress from economic and demographic changes, Canada was admitting far more immigrants in proportion to population; during the 1950s, when immigrants admitted to the United States made up 1.7 percent of its base population, those admitted to Canada equaled

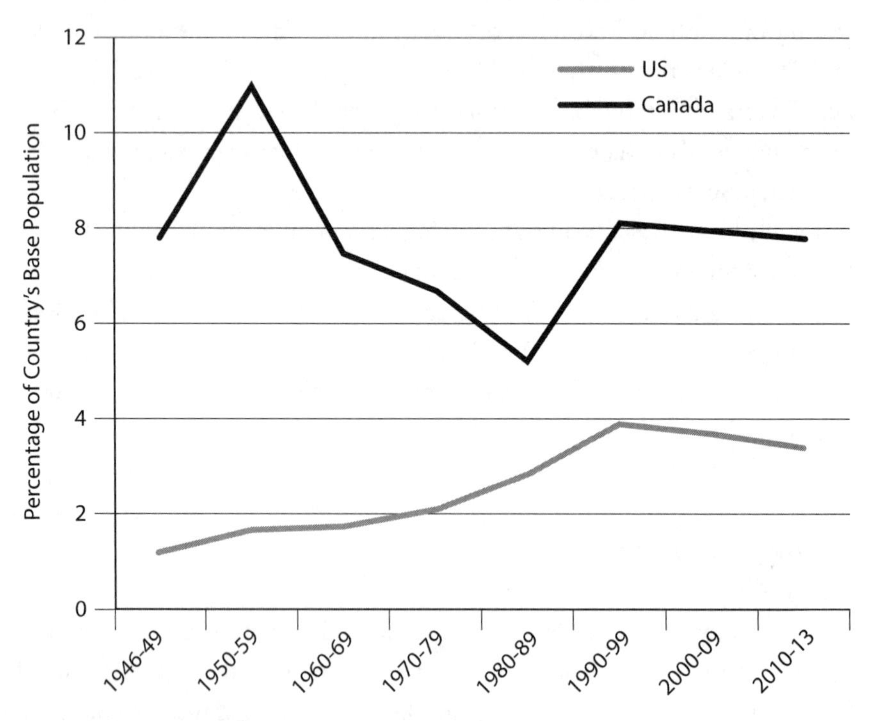

FIGURE 7-2. Permanent immigration to Canada and the United States as a percentage of each country's base population, 1946 to 2013
Source: Compiled from Statistics Canada; US Department of Homeland Security.

11 percent of that country's base population, or more than six times as many. In recent years, the immigration ratio between Canada and the United States in this respect has been fairly steady, at around 2.5 to 1. As we discuss in chapter 8, the great majority of Canadian immigrants settled in Canada's central cities.[19]

As shown in table 7-2, the origins of immigrants to Canada and United States are somewhat different. Canadian immigration is largely from Asia, whereas in the United States it is roughly evenly divided between Asia and Latin America. Immigration between the two countries is modest; in recent years, immigration from the United States to Canada has been running about 10,000 per year, whereas about 20,000 Canadians per year move in the opposite direction.

The origins of immigrants to Canada have shifted significantly over time. Prior to 1971, immigration to Canada was predominately from Europe; during the 1970s and 1980s, as the Asian share of immigration grew, roughly 20 percent of Canadian immigration was from the Caribbean and Latin America, predominately Afro-Caribbean in composition. Since the 1990s, more than half has been from Asia, with a growing share coming from Africa. The principal Asian countries from which immigrants have come to Canada in recent years have been the Philippines, China, and India. By comparison, most

TABLE 7-2.
Place of origin of immigrants, Canada and the United States

Origin	Canada (2006–2011) (%)	United States (2000–2009) (%)
Asia	57	34
Europe	14	13
Africa	13	7
Latin America and Caribbean	12	41
United States	4	—
Canada	—	2
Other, not classified	1	2

Source: Compiled from Statistics Canada 2011 National Household Survey; United States Department of Homeland Security.

immigrants to the United States came from Europe through the mid-1960s, but since the immigration law reforms of that era, roughly 40 percent of immigrants to the United States have come from the Caribbean and Latin America and roughly 30 percent from Asia.

Housing and Urban Renewal Policies

There is less overt difference between housing policies in the United States and Canada than between the United States and many other countries.[20] Both countries have housing markets dominated by private developers and, at least traditionally, private mortgage finance[21] and have small social housing sectors compared with many European countries. Although there have been periods during which the Canadian government exhibited a strong commitment to the production of social housing, particularly during the 1970s, that commitment largely evaporated with the change in political leadership in the early 1990s and has never returned. Although some provinces, notably British Columbia and Quebec, have filled part of the gap, far fewer units of new social housing relative to population size are being built in Canada today than in the United States. This fact does not reflect a strong policy commitment to social housing by the federal government in the United States, which has drastically cut appropriations for social housing, but rather the creativity of policy makers in the 1980s who came up with the Low Income Housing Tax Credit program, a perpetual source of funding for affordable rental housing that exists outside the federal appropriations process. Despite the modest level of social housing production in Canada, fewer Canadians are burdened by housing costs (spending 30 percent or more of their income on housing) than Americans: 25 percent of households compared with 30 percent in the United States.[22] Despite the absence of a home-mortgage interest deduction in Canada, a substantially greater percentage of Canadians are home owners: 69.0 percent in 2011 compared with 64.6 percent of Americans.

Reflecting the dominant role of the provinces in these sectors, the Canadian federal government has no government department engaged with matters of housing or urban affairs. The federal agency with a housing mandate is the Canada Mortgage and Housing Corporation (CMHC), whose principal responsibility, as reflected in its mandate, is to "contribute to the stability of the housing market and financial system"[23] by fostering a stable mortgage market and ensuring a steady volume of private sector housing production. It does so principally through home-mortgage insurance and loan purchasing programs, in essence operating as the Canadian equivalent of both the Federal Housing Administration and Fannie Mae/Freddie Mac in the United States. To the extent that the Canadian federal government appropriates funds for affordable housing, these funds are mainly allocated as block grants to the provinces and territories under bilateral agreements between the CMHC and the individual provinces and territories.[24]

One important difference that is likely to have a direct effect on urban form in the two countries is that mortgage interest is deductible from income tax in the United States but not in Canada. There is strong evidence to suggest that the deductibility of home-mortgage interest not only has a sharply regressive fiscal effect and little or no effect on the homeownership rate, but also that changing the relationship between the "sticker price" of housing and the carrying cost to the owner triggers increased housing consumption. What that means is that the home-mortgage interest deduction encourages people with means to buy more house and land than would be the case without the deduction.[25] Thus, the general tendency of the deduction is to encourage houses to be larger and sit on more land. Economist Richard Voith has analyzed this question and concluded that as a result of this tendency, the home-mortgage deduction encourages sprawl by increasing land consumption while reducing the cost of restrictive zoning, making it more likely that suburban communities will pursue such policies and further constrain the housing options of less affluent households.[26] His conclusion strongly suggests that the tendencies toward sprawl in the form of large-lot exclusionary zoning in the United States, already driven by widespread municipal dependence on local property taxes, are further exacerbated by federal tax policies.

There have been periods, however, when housing policy differences between Canada and the United States have been more pronounced. Although they may no longer be in effect, two such differences should be noted because they still resonate in the difference in urban form we see in Canadian and US cities. The first is the much lower level of urban renewal activity in Canadian cities from the 1950s through the 1970s, a period during which urban renewal in US cities can, not unreasonably, be compared to a scorched-earth policy that leveled thousands of buildings and entire blocks and displaced well over a million households. This activity not only changed the character of most older US urban downtowns, but by trying to impose a quasi-suburban and automobile-dominated model for the rebuilding of these downtowns, we would argue that it impeded—rather than accelerated, as had been its intent—their revival.

The second is the difference in social housing policy during the 1960s and 1970s. Those critical decades saw both countries move from a model based on housing owned and operated by governmental agencies to one in which the private sector was the vehicle for providing social housing. The way in which it was done in the two countries, however, was sharply different. In the United States, the programs of the 1968 Housing Act had the effect of encouraging the for-profit private sector to produce inexpensive, and often shoddily built, rental housing that was then further subsidized by the federal government. In Canada, the government looked to the nonprofit sector to produce social housing, emphasizing permanent affordability, mixed-income development, and cooperative ownership. Although these policies no longer govern what little new social housing is built in Canada today, the social housing inventory created during this period continues to represent a significant and highly valuable share of the nation's urban housing stock.

During the same years, Canadians built two important, outstanding, and even iconic large-scale urban developments: the False Creek project in Vancouver and the St. Lawrence project in Toronto. Both built on underutilized or vacant land, they represent a paradigm of high-density, walkable, and economically integrated urban redevelopment that has served as a successful model for subsequent planning and development in Canadian cities; as a 2014 article described the latter development, it was "the 1970s housing development that to this day remains a shining example of Toronto getting things right. Though it broke all the urban planning orthodoxies of the time—no, *because* it broke them—St. Lawrence achieved the Holy Grail of urban planning: It became a successful, fully-functioning, mixed-use, mixed-income community."[27] Notably, it was influenced by the ideas of Jane Jacobs, who moved to Toronto in 1968 and became, in Christopher Hume's words, "a kind of civic patron [who] gave [Mayor] Crombie, [chief planner] Littlewood and their team the courage to go against the planning verities of the day."[28] One would be hard-pressed to find any comparable examples from the United States during the same period.

Public Education

Another area in which differences between the United States and Canada is likely to have contributed to the relatively greater social and economic strength of Canadian central cities and helped moderate suburban sprawl is the structure of public education. Education is defined as a provincial responsibility in the Canadian constitution; there is no federal department of education as in the United States, and the federal government plays at most a modest role. Many Canadian school districts are organized on a regional basis in which a wide range of social and economic levels and geographic areas are represented, and provinces typically equalize school funding levels to minimize resource imbalances among school districts. As with municipalities, school district boundaries are within the purview of the province; thus, in the 1990s, Nova Scotia consolidated its twenty-one school districts into seven regional districts. This practice is in contrast to that in much of the United States. Although regional school districts are common in rural and not unknown in suburban areas in the United States, single-city school districts with high poverty concentrations and inadequate resources are the rule rather than the exception with respect to central cities.[29] Poor public schools in central cities are often cited as a key factor leading to the decision of middle- and higher-income families to flee the central cities after World War II, a trend that is still continuing in many US cities.

In five Canadian provinces and two territories—Alberta, British Columbia, New Brunswick, Newfoundland and Labrador, Ontario, Yukon, and Nunavut—the entire cost of public education is provided by the province or territory from general revenues. In the others, the lion's share of school costs is borne by the provinces. In Quebec in 2009, for instance, 75 percent of funding for the province's sixty-nine school boards came from provincial sources and only 15 percent from local property taxes, with the balance from other sources, such as federal contributions to special programs, and self-generated revenues.[30] These figures are in marked contrast to the United States, where on average 40 percent of school costs are raised locally and where—despite efforts at reform in recent years—great disparities exist between the resources available and the budgets of low- and high-resource school districts.[31]

Quebec provides funding to school boards based on the number of students in the board's system of schools, so each student in the province is treated equally. School boards are empowered to raise additional revenues through an educational property tax, but the province limits the amount that can be raised in this way and tops up revenues raised by school boards in poorer districts so that all school boards have essentially the same per student revenues from this source. Schools are permitted to raise revenues through other means (selling food in the cafeteria, selling school uniforms, bake sales, etc.), but the amounts involved are usually small compared to the official funding sources. The result is that per student spending by schools is very similar across the province, regardless of the wealth of the neighborhood where the school happens to be located. Moreover, teachers in Quebec are paid on the basis of a standard pay scale that, while taking into account seniority and education, applies universally throughout the province. Thus, there is no financial incentive for teachers to leave schools in lower-income areas to seek higher salaries in wealthier districts.

These factors tend to short-circuit the self-reinforcing cycle that is often encountered in US cities where poorly performing schools in low-income districts struggle with inadequate financial resources and poorly trained staff earning unattractive salaries.[32] It goes without saying that Canadian schools vary in terms of their quality, but it is rare to see the extremes commonly found in US cities.

Taken as a whole, Canadian public schools generally outperform those in the United States in terms of student test scores and retention rates. Results from the 2012 Organization for Economic Cooperation and Development's International Program for Student Assessment (PISA) show consistently better educational outcomes for Canadian students compared with those of the United States, including larger numbers of students performing at high levels and fewer at low levels; for example, 16.4 percent of Canadian students performed at the highest levels in math compared with 8.8 percent of students in the United States, and 13.8 percent performed at the lowest levels compared with 25.8 percent of US students. Canada scored significantly higher with respect to equity in resource allocation, and although more than one of five students in the United States attended socially disadvantaged/low-performing schools, the same was true of less than one of ten Canadian students.[33]

Conclusion

The purpose of this chapter has not been to paint Canada as a socioeconomic paradise, which it clearly is not. Large numbers of Canadians are poor, spend excessive amounts for their shelter, or perform poorly in school. Social and economic conditions for Canada's large Aboriginal population are markedly worse than for the rest of the country's population, much as is the case in the United States. On every important indicator, however, the percentage of Canadians in any form of social or economic distress—whether poverty, housing cost burden, or poor educational performance—is significantly lower than in the United States even though generalized economic data indicate that the United States remains, overall, a wealthier nation.

The principal reason for this variation, we suggest, can be found in the differences in social policy between the United States and Canada. In particular, the focus on equalization of resources and integration of diverse populations that is central to Canadian social policy—from the national policy of fiscal equalization between the provinces and the nation's generous practices in welcoming immigrants to such specific policies as Quebec's province-wide educational pay scales and Nova Scotia's reorganization of public education into seven regional school districts—is crucial. Although Canada does not have explicit social policies designed to reduce economic disparities at the spatial level, such as some of the fair-share housing models in use to greater or lesser extent in the United States, a strong case can be made that Canada's policies that reduce economic disparities as well as the widespread use of municipal amalgamation as discussed earlier powerfully affect the spatial distribution of the population by income, which in turn has been shown to affect the extent of sprawl within a region.[34]

The relationship between the social policies described in this chapter and urban growth and development is nonetheless not always a direct one. These policies, however, have a direct bearing on the greater relative vitality of Canadian central cities and on the notably more modest social and economic disparities between central cities and their suburban surroundings in Canada, as we discuss in the next chapter. Social policies are not the only contributor to these outcomes, but they are significant ones. To the extent that social policies have reduced urban-suburban disparities, they will affect the extent and character of suburban growth as well.

Chapter 8

Vibrant, Diverse Central Cities

Up to now, we have looked at areas where Canadian policies differ today or have differed at critical points in the past from those of the United States. Now our focus turns to how the these policy differences play out in different contexts, beginning with their effect on the character and vitality of central cities in the two countries. This chapter explores how those differences, along with other historical, legal, and cultural factors, have affected the trajectories and present state of central cities in Canada and the United States. Of course, there are many similarities between cities in the two countries as well as differences. In both countries, the condition of individual cities falls on a continuum, and there is more than a little overlap when it comes to population trends, urban form and density, and social and economic trends. At the same time, the typical or normative condition of Canadian cities is significantly different from that of US cities.

Canadian cities and their downtowns have retained far more vitality over the years since the end of World War II than have their US counterparts. Not only downtowns, but other parts of Canadian cities, including the largely single-family residential neighborhoods that make up the greater part of cities in both countries, have far less of the blight and disinvestment that characterize large parts of older US cities. Although some Canadian cities have lost some population over the past decades, largely as a result of shrinking household size, none have seen anything comparable to the massive population loss and property abandonment experienced by many US cities.

In saying that, however, we should stress that we do not mean that *all* Canadian cities are healthier or more vital than *all* cities in the United States. In both countries, cities fall on a continuum from strong to weak, or thriving to distressed. Few cities in Canada or elsewhere are showing more dynamic growth and vitality than Seattle or Washington, DC. At the same time, no city in Canada exhibits the concentrated poverty, abandonment, and distress of Detroit, Cleveland, Buffalo, and a host of other US cities. The central point is that when one looks at the continuum of cities, Canadian cities are skewed to the positive end, and cities in the United States are more skewed to the negative end. The question is why.

Population Trends

In chapter 4, we presented data on population changes in Canadian and US central cities in the post–World War II period (see tables 4-6 and 4-7). Although three of the ten Canadian cities lost population between 1971 and 2011, on average, the group of cities increased population by about 25 percent. This increase reflects a mix of explosive growth in Edmonton, Calgary, and Vancouver with more modest growth elsewhere, including older largely industrial cities like Toronto, Hamilton, and Windsor. Each case in which central-city populations declined can be attributable to the declining size of the typical household. All the Canadian cities saw significant increases in the number of households and housing units, even if they experienced modest population declines. For the five cities for which we have data as far back as 1951, only Winnipeg declined in population over the next sixty years, and the average population growth over that period for the five cities was 28 percent.

Although household sizes were declining all across Canada during those years, the trend was particularly pronounced in Quebec. The average household size in Quebec City dropped from 4.4 persons in 1951 to 2.2 persons in 2011, a 50 percent decline; in Montreal, household size went from 4.0 to 2.3 persons.[1] Winnipeg saw a decline from 3.6 to 2.5, less prounounced than Quebec City and Montreal perhaps, but far greater than the recorded decline in its population during the same period. Thus, even though populations may have been declining in these cities, the actual number of households was continuing to increase.

Chapter 4 also presented data for the ten largest cities in the United States in 1950. Even taking into account modest growth in New York City and significant growth in Los Angeles, these cities on average lost almost 21 percent of their population between 1950 and 2010. With the exception of New York City and Los Angeles, all the cities that were the largest cities in the United States in 1950 lost significant shares of their population by 2010, ranging from 25 percent in Washington, DC, to 63 percent in St. Louis. Although some of these cities, including Washington, DC, and Boston, have begun to regain population in recent years, none is close to returning to its 1950 population, and others continue to lose population.[2] Many smaller US cities—for example, Buffalo; Flint, Michigan; and Youngstown, Ohio—have seen comparable or greater population losses.

Although US cities saw declines in household size during this period, those declines were less pronounced than in Canada, so the effect of this factor on the US cities' population trajectory was less significant. Meanwhile, despite the loss of thousands of households, construction of new housing in the United States continued at a steady pace, contributing to a process of gradual disinvestment and ultimate abandonment of hundreds of thousands of houses and multifamily apartment buildings that continues in many cities to this day. At the same time, deindustrialization and suburban job shifts left comparable numbers of retail, office, and industrial buildings vacant.

All these trends represented a loss of billions of dollars in capital investment, with untold costs in social disorganization and human suffering. Since the 1950s, hundreds of thousands of buildings have been demolished by public agencies at the cost of millions of dollars in public money, and still more millions must be spent to maintain the vacant lots created as municipal tax bases shrink. Cities with an increasingly impoverished and marginalized population, but with fewer and fewer resources to support them, cut back on services while becoming increasingly dependent on state and federal funds. Their citizens are equally dependent on state and federal transfer payments.

That is not the complete picture of cities in the United States, however. New cities in the Sun Belt, such as Houston, Phoenix, and Las Vegas, have captured both population and economic activity from the nation's Northeast and Midwest,[3] reflecting a shift in the nation's geographic center of gravity, a trend that lacks a true Canadian counterpart. Among the older cities, a growing number are showing signs of revival, not only Boston and Washington DC, but more heavily industrial cities like Pittsburgh and Baltimore. The revival of

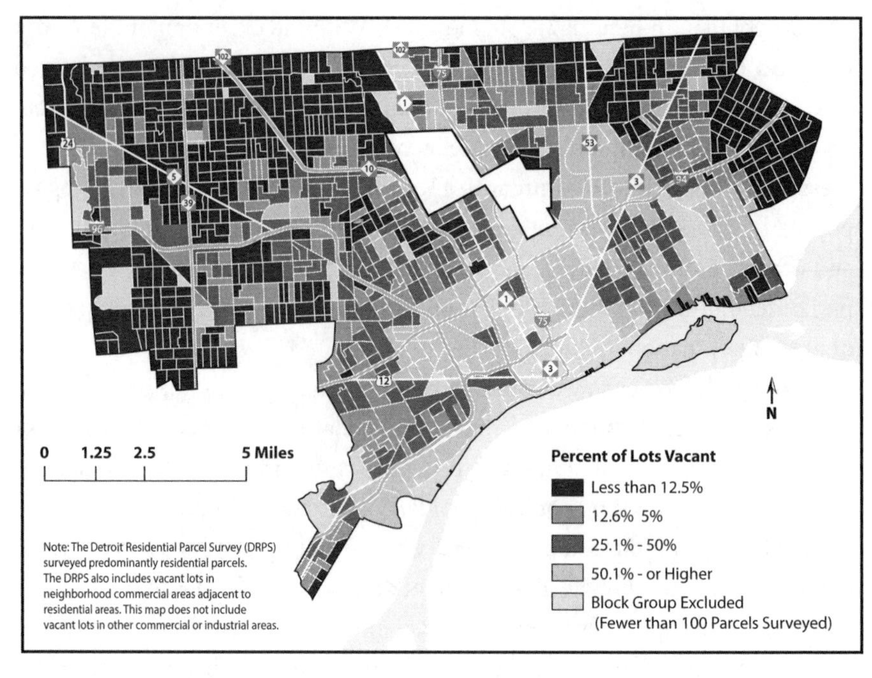

FIGURE 8-1. Vacant land in Detroit, 2009
Source: Data Driven Detroit.

Rust Belt cities like Pittsburgh, however, is fueled heavily by the in-migration of young, well-educated adults into these cities' downtowns and other favored areas; it has led to little improvement in the condition of the rest of the cities or the majority of their populations.

The extent of urban devastation in older US cities is visible in microcosm in Detroit, where it was estimated in 2012 that the city—of a total of roughly 400,000 land parcels—contained 105,000 vacant lots and nearly 50,000 vacant structures, from modest single-family homes to massive iconic structures such as the 1920s Central Michigan railroad terminal. The result has been the landscape shown in figure 8-1 in which large parts of the city have become little more than a wasteland of vacant lots, vacant or derelict structures, and a handful of home owners—usually elderly couples or individuals—scattered across the landscape. If home owners in such areas would like to move, which many would, they would find that their homes have no value on the market.

Demographic Trends

The loss of population in so many US cities is not neutral in its social effects, but reflects the increasing impoverishment and dependency of urban populations relative to their suburban surroundings, something that is much less true in Canada. In contrast to older central cities in the United States, where household and family incomes are generally well below statewide or regional levels and poverty levels significantly higher, household and family incomes in Canadian central cities are generally comparable to or above provincial levels.

Similarly, cities in the United States typically have far fewer married couples with children as a percentage of all households than their Canadian counterparts relative to statewide or province-wide levels. This statistic is significant because these households are typically the most affluent and economically stable of all household types. They represent a critically important element in sustaining a city's economic strength, its social stability, and the vitality of its predominately single-family residential neighborhoods, a neighborhood type that tends to dominate the historical urban landscape outside downtown core areas in most cities in both Canada and the United States.

Comparative statistics for a cluster of Canadian and a cluster of US cities, in both cases showing a mixture of cities from around each country, are shown in table 8-1. Although comparing Canadian and US cities is rendered more complicated by the amalgamation of Canadian cities like Toronto or Halifax into regional entities, the underlying picture transcends the effects of amalgamation. Canadian cities are far more likely to contain populations that are similar to their regions with respect to both income and family type.

Perhaps more significant is that the population of Canadian cities is often significantly more economically integrated in terms of its distribution *within the city* than in cities in the United States. A simple metric has been developed by Sean Reardon and Kendra Bischoff to measure economic integration by looking at how populations distribute within a city relative to the citywide median.[4] When we compared a cluster of cities in the United States and Canada (table 8-2), we found three things. First, a significantly larger share of the Canadian cities' residents than US cities' residents lived in neighborhoods (census tracts) where the median income was close to the citywide median

TABLE 8-1.
Comparative economic and demographic features of selected Canadian and US cities

City	Median Family Income[a]		City % of State/ Province	Percentage of Households That Are Married Couples with Children[b]		City % of State/ Province
	City $	State/ Province $		City	State/ Province	
Regina	70,353	58,563	120.1	25.1	26.4	95.2
Quebec	62,619	58,678	119.4	20.3	25.7	79.0
Winnipeg	62,955	58,816	107.0	25.2	27.6	91.4
Gatineau	70,071	58,678	106.7	28.1	25.7	109.3
Calgary	77,658	73,823	105.2	30.0	30.5	98.4
London	67,018	69,156	96.9	26.6	31.2	85.3
Hamilton	66,810	69,156	96.6	29.3	31.2	93.9
Edmonton	69,214	73,823	93.8	25.4	30.5	83.3
Toronto	59,671	69,156	86.3	25.5	31.2	81.7
Montreal	49,969	58,678	85.2	19.3	25.7	75.1
Atlanta	55,520	58,790	94.4	9.7	22.8	42.5
Minneapolis	60,927	71,307	85.4	12.9	22.0	58.6
Chicago	53,338	68,236	78.2	15.8	22.3	70.9
Dallas	45,162	58,142	77.7	19.3	25.8	74.8
Cincinnati	45,757	59,680	76.7	9.1	19.4	46.9
Boston	58,600	81,165	72.2	10.7	20.7	51.7
Philadelphia	45,619	63,364	72.0	12.4	19.4	63.9
Baltimore	47,435	85,098	55.7	9.7	21.9	44.3

Sources: Compiled from United States 2006–2012 Five-Year American Community Survey and Statistics Canada Census.

[a] Family income was used rather than household income for comparability purposes because cities generally contain a higher percentage of single individuals than states or provinces as a whole and US cities contain a higher percentage of single individuals than Canadian cities. Figures for Canadian cities in Canadian dollars and for US cities in US dollars.

[b] Canadian figures (both for cities and provinces) are slightly higher than US figures because the Canadian census combines married-couple and common-law (unmarried) child-rearing couples, whereas US data are for married couples only.

(between 80 and 120 percent of the citywide median). Second, significantly fewer residents of Canadian cities lived in high-poverty census tracts (below 50 percent of the citywide median) than did their US counterparts. Third, although inequality increased sharply in *all* the US cities between 2000 and 2012, Canadian cities presented a mixed picture in that inequality increased in some but decreased in more.[5]

TABLE 8-2.
Comparative levels of economic segregation and integration in selected cities in the United States and Canada

City	% of Census Tracts with Median Income 80 to 120% of Citywide Median			% of census tracts with Median Income < 50% of Citywide Median		
	2000	**2008–2012**	**Change**	**2000**	**2008–2012**	**Change**
Atlanta	19.0%	18.4%	– 0.6%	19.8%	21.3%	+ 1.5%
Baltimore	37.2	31.8	– 5.4	7.5	10.1	+ 2.6
Denver	47.4	34.0	–14.4	3.0	2.8	– 0.2
Minneapolis	42.5	36.5	– 6.0	6.7	13.0	+ 6.3
St. Louis	45.0	27.7	–17.3	5.4	6.9	+ 1.5
	2001	**2011**	**Change**	**2001**	**2011**	**Change**
Halifax	56.8%	39.3%	–17.5%	1.6%	2.2%	+ 0.6%
Hamilton	35.4	33.3	– 2.1	5.4	4.1	– 1.3
Saskatoon	42.9	57.8	+14.9	2.4	0.0	– 2.4
Vancouver	55.2	58.6	+ 3.4	3.8	2.6	– 1.4
Winnipeg	41.1	43.8	+ 2.7	6.0	4.6	– 1.4

Sources: Compiled from US Census, 2008–2012 American Community Survey; and Statistics Canada 2001 and 2011 Census.

As we can see in table 8-2, not only in Saskatoon did *no* households live in high-poverty census tracts in 2011, but the same was true for a number of other Canadian cities, including Kitchener and Quebec City. By contrast, more than one out of every five residents of Atlanta live in a high-poverty census tract. What is even more notable, however, is that a number of the US cities—Atlanta being the notable exception—showed distributions not markedly different from those of Canadian cities as recently as 2000, but have diverged sharply since then. Today, many Canadian cities are not only more vital than many US cities, but more economically integrated as well.

Why Are Canadian Cities Different from US Cities?

As with any other complex system, there is no one reason to explain the greater continued vitality of Canada's urban centers compared with the older cities of the United States. Although many of the same factors that have led to differences in suburban growth patterns also bear on the greater continued vitality of Canada's urban centers, other factors, such as immigration as well as social and economic factors that may not be within anyone's ability to control, have had a particularly strong effect on the nation's central cities. Let us untangle some of the principal threads.

One factor is that provincial policies have enabled Canadian cities to steadily expand their boundaries through consolidation with their suburban neighbors, even when opposed by some of the cities or suburban municipalities involved. By increasing the elasticity of Canadian cities, they have fostered greater social and economic diversity within the expanded city and encouraged stronger metropolitan growth; as David Rusk points out, the smaller the income gap between city and suburb, the greater the economic progress for the entire metropolitan area.[6] Chris Benner and Manuel Pastor have made a similarly strong case that the less the economic polarization in a region, the stronger its climate for economic growth.[7]

Another factor is the greater extent to which provinces provide resources to equalize fiscal capacity among local jurisdictions, although it would not be appropriate to give this factor much weight. Canadian cities lack many of the fiscal tools available to many, although not all, US cities. Canadian cities cannot levy sales or wage taxes, and they are far more dependent on local property taxes than US cities.[8] Their level of fiscal stress and the fiscal disparities among them, however, are both less pronounced than in the United States. Although Canadian cities face fiscal difficulties and constraints in the provision of municipal services, few if any suffer from the level of fiscal crisis prompting a small but growing number of US cities to take refuge in bankruptcy.[9]

This picture is in marked contrast to that of the United States. A handful of US central cities—for example, Indianapolis, Nashville, and Louisville—have consolidated with their suburbs into single regional entities over the past few decades; these consolidations are not only rare in the extreme, but have all been purely voluntary. Although American states have the power on paper to compel municipalities to consolidate, as we discussed earlier, no state has or in all likelihood would seriously entertain exercising that power. A handful of central cities continue to grow through annexation, but such practices are growing rarer; moreover, the cities that are able to do so are almost entirely the "new" cities of the Sun Belt. Elsewhere, the landlocked condition of US central cities reinforces the disparities with the growing suburban areas around them, in contrast to Canada.

Transportation policies are critically important to central cities as well as suburbs. Fostering transit, maintaining higher gas prices, and constructing fewer miles of freeways all discourage centripetal effects and sustain the vitality of central areas. Although residential suburbanization in the United States began well before the arrival of the interstate highway system, most of which typically did not come on line until the second half of the 1960s, it was the proliferation of highways and the creation of high-traffic highway nodes that accelerated the suburbanization of nonresidential functions and the creation of the vast number of suburban shopping malls and office parks that make up such a large part of today's US suburban landscape.

Recent Canadian governments have shown little interest in urban policy as such, but many of the key policies that have reinforced today's differences between US and Canadian cities took place many years ago. Indeed, to seek an inflection point in the differing trajectories of cities in the two countries, it is productive to go back to the 1960s and 1970s and look at the continuing effect of policies and actions dating from those years, as discussed in chapter 7. They were critical years for United States cities, years in which a gradual decline that followed the end of World War II turned, in many cases, into a precipitous downward slide.

Urban renewal had far less effect on Canadian cities than on their US counterparts. Although Canadian cities were not immune to urban renewal in the "US style" during the 1960s and 1970s, Canadian cities saw far less of the large-scale removal of their nineteenth- and early twentieth-century fabric than did US cities during that period. Significantly, no Canadian national government during that period ever created a program, let alone provided billions of dollars behind it, explicitly designed to further the removal of cities' historic urban fabric.

Moreover, during that same period, a number of significant developments took place under the rubric of urban renewal in Canada that represented significant progress toward creative, sustainable urban vitality rather than undermine the cities' urban fabric. The False Creek South project in Vancouver[10] and the St. Lawrence project in Toronto were and are still models of high-density, walkable, and economically integrated urban redevelopment, combining home ownership and rental, subsidized, and market-rate housing with nonresidential uses and public open space.

One would be hard-pressed to find comparable examples from the United States during that period. Urban housing development in the United States at the time was little but large, single-use, means-tested developments built with federal subsidies, which were available in generous amounts in the 1970s. Although many of these developments were well designed and are well managed, their larger social effect was to increase poverty concentration in urban areas and further exacerbate already large urban-suburban imbalances. Largely as a result of exclusionary zoning and other practices, as well as the inability of the cities and the unwillingness of most states to influence suburban zoning practices, few subsidized housing developments for lower-income families were built in the suburbs of US cities during those years or since.

In retrospect, it is now painfully clear that the urban renewal of US cities—with few exceptions—did little to enhance their short-term prosperity, and by reducing their walkability and historic character, became a major impediment to future revitalization. It is telling that the revival of downtown St. Louis has been led by Washington Avenue and that of Cleveland by the Warehouse District, both areas at the downtown fringe that largely escaped the bulldozer that leveled much of the rest of those cities' downtowns.

Canadian downtowns have remained the hearts of their regional economies to a much greater extent than in the United States. They typically contain not only a large part of their regions' office space, as is also true of many US cities, but continue to function as major regional retail centers. In recent years, they have significantly increased their residential population as well. Although the downtowns of many US cities are reviving, including those of some otherwise deeply distressed cities like St. Louis or Baltimore, only a few, such as Chicago or San Francisco, have retained a significant role as regional general-purpose retail and service centers.[11] Others, such as Detroit or Buffalo, have largely lost those functions and continue to contain large numbers of abandoned buildings and vacant lots.

Social policy differences between Canada and the United States are also relevant to the state of the cities. As discussed in chapter 7, Canada's stronger social safety net reduces income inequality in Canada; although it is greater than in many European countries and increasing, it is still far less in Canada than in the United States. Greater income inequality not only works to undermine central-city vitality in itself, but the inequality between central cities and suburbs in most parts of the United States, particularly in the older regions of the Northeast and Midwest, is arguably a major factor blocking consolidation of cities and suburbs and further encouraging exclusionary zoning in suburban jurisdictions. Moreover, for all the rhetoric about the United States being a nation of immigrants, Canadian cities, led by Toronto and Vancouver, have accommodated large numbers of highly diverse immigrant communities with more apparent success than many US cities.

The significance of a more generous immigration policy to the vitality of Canadian cities should not be underestimated. As discussed in chapter 7, Canada admits more than double the number of immigrants as a percentage of its national population than does the United States. Of particular significance for urban trajectories is that during the critical years of the 1960s and 1970s, when US cities were experiencing their greatest levels of population loss, the disparity was far greater. Between 1965 and 1974, immigrants admitted to Canada represented 8 percent of the nation's 1970 population, whereas immigrants to the United States during the same period made up only 1.8 percent of the US 1970 population. A substantial majority of Canadian immigrants settled in the nation's central cities.

Immigrants have played an important role in the health of Canadian central cities, although, as is true as well in the United States, more in some than in others. In 1981, for example, 52 percent of all immigrants to Canada lived in the Toronto, Vancouver, and Montreal areas, home to 26 percent of the nation's Canadian-born population; by 2001, those three areas' share of Canada's immigrant population had risen to 62 percent, and by 2011, it had increased to slightly more than 63 percent. Table 8-3 shows the percentage of immigrants by city for Canada's major cities in 2006; although the percentages vary widely, it remains that more than 20 percent of the population of seven of the eleven cities is made up of immigrants. Although there are some cities in the United States with large immigrant populations, there are few outside the Southwest, where most major cities have large Latino populations. Cities like Minneapolis, Seattle, and Washington, DC, are widely seen as diverse, cosmopolitan cities, yet in none is as much as 20 percent of the population made up of immigrants.[12]

TABLE 8-3.
Percentage immigrant population by city, 2006

City	% Immigrant	City	% Immigrant
Toronto	50.0	Ottawa	22.3
Vancouver	45.6	Winnipeg	18.7
Montreal	30.7	Regina	8.0
Hamilton	25.4	Halifax	7.4
Calgary	24.8	Quebec City	4.4
Edmonton	22.9		

Source: Compiled from Statistics Canada Census.

In New Jersey, Newark's Ironbound neighborhood, which was revived by immigrants from Portugal and the Cape Verde Islands in the 1950s and 1960s, is the city's highest-value residential area and a destination for Brazilian and other Latin American immigrants. Its commercial strip along Ferry Street, which draws thousands of shoppers and visitors every weekend, is a major economic engine and revenue generator for the city. In recent years, southwest Detroit has been revived by Mexican immigrants and nearby Dearborn by immigrants from Lebanon, Palestine, and other Arab countries. One can only speculate on how different the fate of many cities in the United States might have been if they could have had access to a larger pool of immigrants to replenish their population and generate economic activity during those critical years of middle-class flight and job loss.

On a related theme, Canada has been fortunate to have escaped much of the corrosive effect of the history of white/black racial conflict that continues to afflict the United States. That conflict not only led to the riots that convulsed US cities in the 1960s and 1970s, with devastating effects on those cities' social and economic condition, but has more broadly created a legacy of both realities and perceptions that continues to affect these cities deeply, including acting as a further factor exacerbating suburban exclusion. Although it can reasonably be argued that—given the importance of suburbanization, regional shifts in population to the Sun Belt, and massive loss of manufacturing facilities and jobs—racial tensions and conflicts did not drive the decline of older US cities during the second half of the twentieth century, it is equally reasonable to argue that those conflicts significantly exacerbated the decline and rendered the subsequent recovery that much more difficult.

Crime is another factor that differentiates cities in the United States and Canada. Abundant research has established links between crime and disorder on the one hand and out-migration and increased concentrations of poverty in urban neighborhoods on the other.[13] Crime rates in Canada are generally significantly lower than in the United States; although Canada has 11 percent of the population of the United States, the number of murders in the country in 2011 was 598, or 4 percent of the 14,612 murders in the United States during the same year, which was not much more than the average number of murders in Chicago or Detroit in recent years. Although some Canadian cities have higher crime rates than the Canadian national average, there is no overall tendency for cities to be high-crime areas as is true in the United States.

Moreover, even Canadian cities that are seen as dangerous from a Canadian perspective are safe by comparison to US cities. Winnipeg, which has been characterized as the most dangerous of Canada's cities of more than 250,000 people,[14] had 2011 homicide and robbery rates that were both significantly lower than the US average. Of Canada's one hundred largest cities, more than half—Calgary, Toronto, Gatineau, Quebec, and Ottawa, among others—had crime rates *below* the national average. This statistic is in contrast to large US cities, which, with few exceptions, have crime rates well above the US national average. Because one cannot compare overall violent crime rates between the two countries due to differences in definitions, data on just homicide rates for US and Canadian cities appear in table 8-4.

Why crime rates are so much lower in Canada is obviously a complicated matter. The reason is likely to be in some large part the product by cultural and social factors resistant to public policy intervention. At the same time, public policy plays an important role. Differences in poverty rates and inequality between cities in the United States and Canada are likely to be relevant, as may be Canada's substantially more restrictive gun control laws, which are likely to significantly reduce the availability and use of firearms.

One last area that is likely to make an important contribution to the social and economic strength of Canadian cities is the structure of public education. Although the United States may not want or need to emulate the Canadian pattern of dual English and French language school districts (or the anomaly, from

TABLE 8-4.
Comparative homicide rates for selected Canadian and US cities, 2011

Canadian Cities		US Cities	
City	Homicides per 100,000 Population	City	Homicides per 100,000 Population
Regina	4.0	Detroit	48.3
Winnipeg	3.2	Baltimore	31.3
Edmonton	3.1	Philadelphia	21.2
Hamilton	2.3	Atlanta	20.7
Toronto	2.2	Cincinnati	20.5
Montreal	1.9	Chicago	15.9
London	1.9	Dallas	10.9
Calgary	1.3	Boston	10.1
Quebec	1.1	Minneapolis	8.3
Gatineau	0.4	San Francisco	6.1

Source: Data from Statistics Canada compiled and published by Taylor-Vaisey 2012.

a US perspective, of public Catholic schools), the Canadian experience offers useful insights that may help explain the consistently stronger performance of Canadian public schools compared with their US counterparts. Among these factors, discussed in chapter 7, are the regionalized nature of most Canadian school districts and the equalization of fiscal resources between school districts at the provincial level. Canadian public schools generally outperform those in the United States; given the relatively low performance of urban public schools in the United States compared with the US national average, it is likely that the disparity between urban public schools in Canada and the United States is significant.

Many reasons have been offered for the better performance of Canadian public schools, A 2011 *Education Week* blog post[15] on this cited a number of possible reasons:

- Better trained teachers, reasonably well paid, with good job security and unionization. It is hard to get into teaching in Canada, but teachers there are generally respected and treated well.

- A strong commitment across the country to equity for all population groups (although there are still large achievement gaps in Canada, they are smaller than in most other countries).
- Better basic services for all students and families, such as health care and social services generally.
- Much smaller differences in funding levels from one district to another and generally more spending in higher need communities.
- Much consistency across schools and districts in curriculum and teaching methods.

Although a detailed examination of these issues or the most appropriate policy responses to them is beyond the scope of this book, finding ways to improve US inner-city education outcomes continues to be a critical issue for restoring vitality to US cities.

Comparing Two Cities: New Haven and Halifax

Many of the themes discussed above in general terms can be seen in a comparison of two specific cities and their trajectories during recent decades. For this comparison, we have selected New Haven, Connecticut, in the United States and Halifax, Nova Scotia, in Canada, two smaller cities that are among the oldest cities in their respective countries.

After St. John's, Newfoundland, Halifax is the oldest city in Anglophone Canada, founded in 1749 on the east side of a 10-square-mile peninsula as a British fortress and naval base. Over the next two centuries, the region gradually developed on a base of resource extraction, fisheries, and agriculture, with most of its wealth drawn to Halifax as the provincial administrative, financial, and port center. New Haven, which was founded by religious dissidents in 1638, was one of the major centers of America's first industrial revolution at the end of the eighteenth century. It remained a major industrial city through the middle of the twentieth century. Industry in Halifax, including textiles and shipbuilding, was on a more modest scale; ironically, today 14 percent of the jobs in Halifax are in manufacturing compared with less than 4 percent in New Haven. Both cities, however, went through significant and traumatic change in the second half of the twentieth century.

New Haven was one of the first cities in the United States to anticipate the decline associated with suburbanization and deindustrialization after World War II, and under the leadership of Mayor Richard E. Lee (1954–1970), New Haven became nationally known as a pioneer in urban redevelopment and antipoverty strategies. Using millions in federal urban renewal funds, large parts of the city's downtown and inner neighborhoods were leveled in the 1950s and 1960s and replaced with office buildings, a downtown mall, and subsidized housing, but displacing more than one out of every seven city households.[16] During the same years, some fifty lane-miles of freeway were carved out of the city, with Interstate 95 cutting across from east to west and connecting to northward Interstate 91, and Route 34, the so-called downtown connector, carved out of the edge of downtown and subsequently dubbed the "highway to nowhere."[17]

Although it is impossible to tell precisely what might have happened in the absence of this massive effort, characterized by one writer as "a bold program that sought to save the city by remaking it for the automobile,"[18] it does not appear to have reversed, or even slowed down, the city's decline. In the 1970s and 1980s, New Haven went into free fall. Riots racked the city in 1967, and between 1950 and 1980, the city lost nearly one-fourth of its population. New Haven's historic manufacturing base effectively disappeared, and crime skyrocketed. Once one of the nation's safest cities, New Haven in the 1980s became one of the most dangerous (figure 8-2). New Haven arguably hit bottom around 1990.

Halifax went through a similar, although less extensive transformation. Following the 1957 Stephenson report[19] calling for large-scale redevelopment, Halifax embarked on a series of urban renewal projects in older areas in and near downtown and the harbor. The demolition of Africville, an area that had been a center of the city's black population since the early nineteenth century, was particularly controversial. The project was initially welcomed by large parts of the black population as a salutary effort to replace an area described as

> *almost entirely bypassed by any modernizing influence. Ranging from sturdy but modest bungalows to "rude shacks made of tin sheets and boards, held together by tarpaper and paint," the salt air had caused*

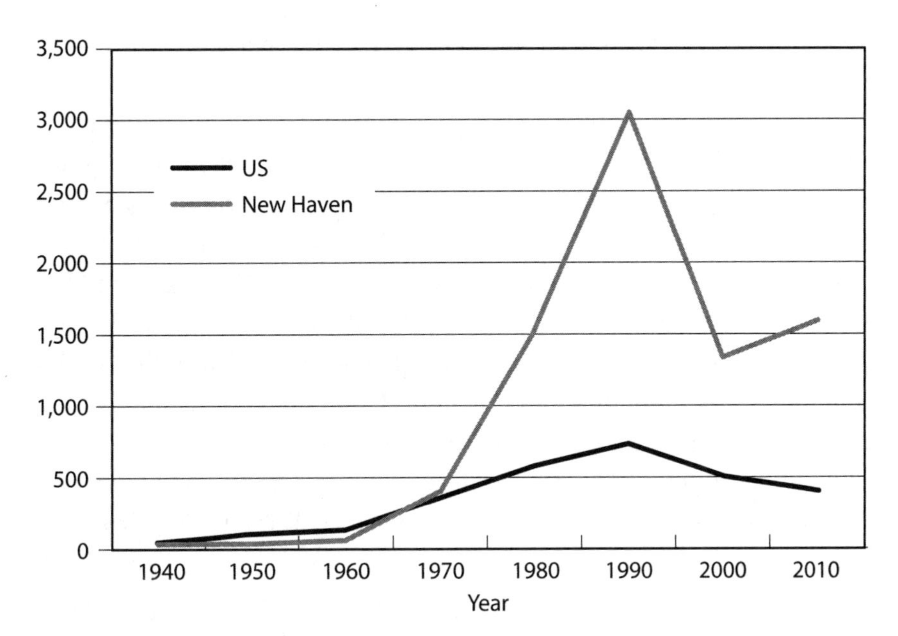

FIGURE 8-2. Violent crimes per 100,000 population in New Haven and the United States, 1940–2010
Sources: Compiled from Rae 2003; FBI Uniform Crime Reports.

> *paint to peel leaving the houses mottled yellows, blues, reds and greens. Piped sewage and water were unavailable; electricity was absent; the sole road was unpaved.*[20]

A combination of delays, bureaucratic snafus, and changing attitudes led to increasing opposition during the 1960s, and the project has been widely seen since as an almost archetypal example of high-handed government and racial insensitivity. Another project, with uncanny parallels to New Haven, was the Cogswell Interchange, a highway project similar to New Haven's and likewise dubbed Halifax's "road to nowhere," which destroyed a large swath of old warehouses near the harbor. A reaction to urban renewal set in by the early 1970s, led in part by a growing historic preservation movement, which ended this phase of the city's history. Since the 1970s, population within the city's preamalgamation boundaries has remained fairly stable.[21]

From this point onward, the paths of the two cities began to diverge. Halifax became increasingly interconnected with its region, from a planning and eventually political standpoint, whereas New Haven's course was largely independent of significant regional connections.

Suburbanization in the Halifax region increased in the 1950s and 1960s. Anticipating the need to provide services and infrastructure to a growing suburban population, discussion of regional planning frameworks began in the early 1960s, leading to the 1969 Provincial Planning Act creating the Metropolitan Area Planning Committee (MAPC). The MAPC adopted the Halifax-Dartmouth Regional Development Plan in 1975, which proposed land use planning strategies and a regional growth boundary along with the creation of regional parks. The powers of the MAPC were more limited than those of similar entities being created elsewhere in Canada during this period. As a result, the regional planning approach soon came undone from pressures that reflected conflict with customary patronage policies, territoriality and a reluctance to share municipal autonomy, the conservatism of the Nova Scotia population, and a distrust of regional planning as something being promoted from outside.[22] This period nonetheless left the Halifax region with a much improved regional infrastructure, which has had considerable influence in shaping the central portion of the Halifax region ever since.[23]

The 1980s and early 1990s were years of economic stagnation in the Halifax area, and much of the renewed impetus for regional planning in the 1990s came from the fiscal crisis of its local governments and the need to spur economic development.[24] Creation of a single-tiered regional government through amalgamation of all the jurisdictions in Halifax County into a single regional municipality was recommended by a provincial task force in 1992 and was adopted after the then-premier, a former mayor of Halifax's neighboring city of Dartmouth, reversed his opposition to the measure. Although the announcement, in the words of one writer, "was met with considerable disapproval by all four mayors and the public in the Halifax-Dartmouth region,"[25] it became law in 1995 and took effect in 1996. The new entity is known as the Halifax Regional Municipality, or HRM. In 2006, the HRM adopted a new body of regional planning policies, stressing the importance of the regional core, designating 25 percent of new housing to be located in the core, and guiding growth outside the core into a series of mixed-use centers located in areas of established settlement and where provision of sewer and water service were most feasible. Since then, master-planned areas in the HRM offering a variety of housing types at higher densities with integrated commercial uses have grown steadily, and regional public transit service has expanded, including bus rapid transit from suburban areas to core employment nodes. In 2013, of all trips in the HRM 12.5 percent were by public transit compared with 4 percent in the New Haven metropolitan area.

Since 1990, New Haven has seen a remarkable turnaround, going from "being an Ivy League punch line to a place where well-heeled Shoreline suburbanites come for a dose of urban glamour."[26] Many different factors have gone into the city's revival. The enlightened leadership of Mayor John DeStefano from 1994 to 2014 played a role, but arguably more important was the belated recognition by Yale University that, as not only the city's dominant employer but its sole significant economic engine, its future was inevitably dependent on New Haven's vitality. Although the relationship between Yale and the city had historically been adversarial, under President Richard Levin's leadership, Yale has become a major partner for the city in redevelopment, neighborhood revitalization, and education since the 1990s. Since 1994, Yale has provided more than $25 million to enable more than one thousand of its employees to buy homes in New Haven and spends more than $4 million per year to underwrite higher education for the graduates of New Haven's high schools.

Redevelopment efforts have been revived, but they are radically changed from the urban renewal era. New Haven, with Yale as its partner, has focused on maintaining the remaining fabric of the city, filling vacant buildings—more often than not with upscale apartments and condos—populating streets, and reviving neighborhoods. Many redevelopment efforts have involved reversing or repurposing products of the urban renewal era. Many signature buildings of that era, like the two downtown department stores or the New Haven Coliseum, have been razed, and at long last, plans are under way to remove the Route 34 connector and replace it with mixed-use development. It is hard to escape the conclusion that in today's city, the history of urban renewal and highway construction from the 1950s through the 1970s is a burden to be overcome, not a legacy on which to build.

Since 1995, a school construction program has replaced or restored the lion's share of New Haven's public schools, creating facilities that "are designed to enhance the learning environment for students and staff alike ... and provide facilities that can be used year-round by the school and community."[27] Thanks largely to a supportive state government, which picks up nearly 60 percent of the total cost, New Haven's public schools are relatively well funded, with per pupil school spending 40 percent higher than the statewide average.[28] Parts of the city are humming with a vitality that they have lacked for decades, and between 2002 and 2011, the number of jobs in the city increased by nearly 10 percent.

The picture in Halifax in 1995, as amalgamation was taking place, was also difficult. Cuts in defense spending had hit Halifax particularly hard; as a municipal report put it, "the economic scenario in the region was bleak.... Nova Scotia ranked last in employment growth in Atlantic Canada, and consumers were worried about unemployment."[29] Amalgamation offered an opportunity for new and creative thinking, which led in turn to the creation of the Greater Halifax Partnership as the HRM's economic development arm. The partnership set itself highly specific goals:

- Grow the economy. Support the creation of 20,000 new jobs and reduce unemployment.
- Build and leverage a brand appropriate for the city.
- Reduce the cost of economic development to municipal taxpayers.
- Attract and retain at least $1 million in local private-sector investment in support of the partnership and its economic growth initiatives.

By 2010, the partnership could report significant success. The number of jobs in the region grew from 164,000 to 216,000, and the unemployment rate dropped from 8.7 percent to 6.6 percent, below the national average. Over the same period, municipal funding for economic development was reduced from $4 million in the year before creation of the partnership to a 2010 level of $1.4 million.[30]

Downtown Halifax faced significant challenges. Although urban renewal had left scars, the downtown overall retained much of its historic character, conveying a rich sense of place to residents and visitors. Creating an environment that would both support development and investment while balancing the many agendas of developers, businesses, and an increasingly culturally and socially diverse population was a challenge. The 2009 HRMbyDesign plan for downtown was designed to address the challenge. It provided for mixed land uses, form-based regulations to control massing, and design guidelines to control appearance.

Construction on the Halifax peninsula has accelerated in recent years, including substantial numbers of residential and mixed-use developments, after the 2008 economic slowdown. Although some growth may be attributable to the effects of the downtown plan, the 2011 announcement of Irving Shipbuilding's successful bid for a federal shipbuilding contract potentially worth $25 billion has had a major effect on downtown. Significantly, in 2014, the HRM approved a plan to remove the Cogswell Interchange and use the reclaimed land for predominately residential mixed-use development and public open space.[31]

The residential market is strong, and during the first quarter of 2015, sales prices in the former Halifax City were 26 percent higher than those of the HRM as a whole. Housing production has shifted strongly to multifamily housing from single-family and semidetached homes. That is very different from New Haven, where little new housing is being built in the central city and where total housing production in New Haven County, an area of more than double the population of the HRM within commuting distance of New York City, has averaged less than half that of the HRM since 2005. In contrast to the HRM, where 53 percent of housing production has been multifamily housing—with a marked increase in the share since 2010—only 29 percent of housing production in New Haven County has been multifamily housing (figure 8-3).

Halifax takes pride today in its high quality of life and natural setting. Its economy is highly diverse; although defense spending is still significant, Halifax is a regional center for higher education, health care, and public services, and trade and manufacturing are also significant employment sectors. The median family income in 2011 was $80,490, 8 percent higher than the national median and higher than that of any other population center in Canada's Maritime provinces.[32] By contrast, the median family income in New Haven in 2012 was $46,145, well below that of Halifax even after adjusting for the difference in US and Canadian dollar values and nearly 30 percent below the national median.

Halifax's resurgence is not without troubling issues. Affordability on the peninsula, exacerbated by the limited land supply and increasing gentrification, is

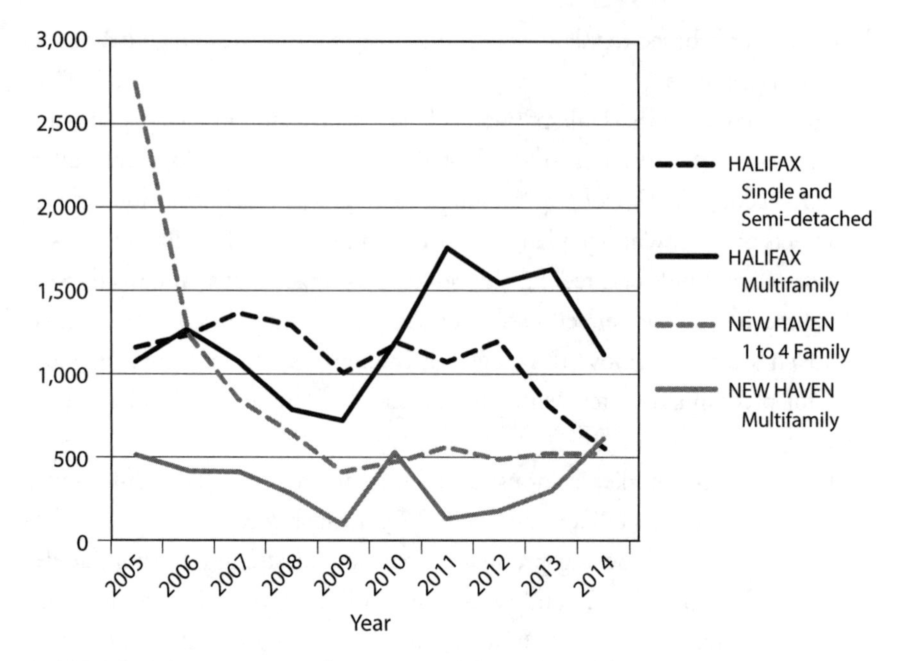

FIGURE 8-3. Housing production in Halifax Regional Municipality and New Haven County, 2005–2014
Sources: Compiled from Canada Mortgage and Housing Corporation; US Bureau of the Census.

a growing concern. Pockets of poverty are emerging in suburban areas such as North Dartmouth, and crime rates are well above the national average. Moreover, in a situation not unlike that of Boston and the rest of Massachusetts, Halifax's growth has been almost entirely at the expense of the more rural areas of Nova Scotia, which are seeing their economic activity wither as Halifax absorbs an increasing share of regional growth. Halifax needs also to consider the future prospect that rather than attempting to manage growth it may have to address shrinkage. Population growth has declined to a trickle, and the population is aging. Immigration to Halifax is below the national average, not only less than immigrant magnets like Toronto and Vancouver, but also less than secondary cities like Winnipeg, Hamilton, and Windsor. In the past few years, attracting more immigrants has risen high on the agenda of both public and private organizations in Nova Scotia. Halifax is also turning to creative city strategies, an effort marked by visits from creative-city gurus Richard Florida in 2004 and Charles Landry in 2009, and is trying to build on the city's vibrant music scene.

The New Haven story, although in part one of resurgence, contains a darker subtext. Despite the revival that started in the 1990s, New Haven remains today a poor city, with a median income only 60 percent that of its region and where 27 percent of all the city's people live below the poverty level compared with 16 percent below the Canadian low-income cut-off in Halifax. Thesse statistics reflect New Haven's continuing role as the home of the region's poor and people of color. New Haven has a large Latino population, which grew from 10,000 in 1980 to 36,000 in 2010, roughly 27 percent of the city's population, but it is not a major immigrant destination. More than half of the city's Latino population is Puerto Rican, and many others have come as secondary migrants from other US cities. Compared to a peer group of ten similarly sized New England cities, New Haven had the fourth highest share of people in poverty and, even more troubling, the second highest Gini index, a measure of income inequality. Although the median income of non-Latino white households has grown by 42 percent since 2000, that of African-Americans has grown by only 22 percent, much less than the rate of inflation, and of Latinos by 13 percent.[33] Although the city has seen job growth, the jobs are far more likely to be filled by suburban commuters than by city residents; in 2011, of all the jobs in the city, 77 percent were held by commuters.[34]

New Haven's region has considerable wealth, but it is largely concentrated in a handful of suburban towns. The South Central Region (a state designation), of which New Haven is the center, contains fifteen separate municipalities with a total population of 570,000.[35] No regional governance exists. Connecticut abolished its counties in 1960 and put nothing in place to replace them. Although a regional council of governments exists, its role with respect to planning and housing is purely advisory; in its own words, it "provides a platform for inter-municipal coordination, cooperation, and decision making."[36] Each town or city in the region sets its own agenda. New Haven, as a small, largely developed city with a disproportionately lower-income population, is, in contrast to the historic city of Halifax, on its own. Although Yale

University's presence and growth have fueled the city's revival, New Haven has become all but a de facto company town, with all the risks of single-industry dependence associated with that status.

New Haven's story is a microcosm of the US urban experience since the end of World War II. The years of urban free fall, when politicians and pundits despaired of the future of the cities, are behind us, yet the future may be holding something almost as troubling: an era of cities polarized spatially, economically, and racially. In this respect, Halifax, for all its complexities, appears in a more positive light in which the regional framework for governance and decision making, coupled with a more diversified local economy, may be putting the city on a course for a more sustainable future.

Conclusion

Canada today may not be pursuing focused, intentional strategies to promote strong, vital urban centers, but just the same, a variety of Canadian policies, both historic and those currently in effect, contribute to that outcome. Establishing the effect of public policies on the vitality of Canada's urban centers is difficult; unlike the effect of land use planning on suburban growth—where one can point to explicit statutes, regulations, and plans—Canada has little in the way of explicit urban policies. Some of the same factors that positively affect suburban development have also played an important role in maintaining urban vitality. In addition to greater transit use, higher gasoline prices, and less highway mileage, all of which help sustain the central functions of the cities, provincial policies mandating the consolidation of cities into larger regional entities have fostered greater social and economic integration and reduced urban-suburban imbalances.

Three historic factors have played a major role in leading to the greater vitality of Canadian cities compared with their US counterparts. First, Canadian cities largely escaped two forces that affected US cities and that have been long-term impediments to revival: most Canadian cities escaped the worst effects of urban renewal, which decimated the cores of US cities during the 1950s and 1960s. Second, Canadian cities have been subject to far less of the white/black racial conflict that has been such a powerful subtext in the post–World

War II story of US cities. Third, the significantly higher level of immigration to Canada, particularly during the 1960s and 1970s, provided an important injection of energy and human capital into Canadian cities at a critical moment in their history.

Two contemporary factors in maintaining vitality in Canada's cities today are the cities' low crime rates and generally successful public education systems. Although hard to pin down, the role of both factors in maintaining the vitality of Canadian cities is likely to be significant. Although one can only speculate about the reasons for the markedly lower crime rates in Canadian cities, reasons for the success of Canadian urban public education include the regional and economically diverse character of most public school districts, strong provincial fiscal equalization measures, and other features of Canadian public education that have rendered it generally more successful regarding pupil outcomes—urban or otherwise—than its US counterpart.

In conclusion, although acknowledging strong cultural, social, and political differences between Canada and the United States, we have made clear in this chapter that the differences in the trajectory of central cities in the two countries are not simply a cultural artifact. Instead, they are rooted in identifiable, explicit differences in policy and practice between the two countries.

Chapter 9
Growing Sustainable Suburbs

In the United States, people routinely make a distinction between suburb and city, seeing the two as almost polar opposites. Such distinctions are blurrier in Canada. Although Canadians clearly recognize the fundamental spatial difference between central cities and suburban areas, the term *city* is often used to describe the larger area, reflecting that consolidation has made many Canadian cities like Halifax, Winnipeg, and Calgary all but coterminous with their regions while adding large swaths of historically "suburban" districts to cities like Toronto and Ottawa.

Canadian municipal boundaries are far more fluid than those in the United States; as Martin Turcotte writes, "Boundaries can change abruptly at any time.... Neighborhoods and localities that had long been considered suburbs can suddenly become part of the central municipality, even though there has been no substantive change in their areas' nature or their social and economic ties to the center."[1] Although large parts of cities like Ottawa or Calgary are suburban in character, Vancouver is the only major Canadian city located in a region where the population is dominated by legally distinct, separate suburban communities, in marked contrast to the United States, where central cities typically contain only a small part of the regional population (figure 9-1).

Social and economic variations between cities and their suburbs are less pronounced in Canada than in the United States, thus further reducing the impetus to make hard-and-fast distinctions between suburb and central city. In the United States, in contrast, many suburbanites' identity is often less about their suburb itself than about their distinguishing themselves from the residents of the presumably impoverished and dangerous central city. Although that tendency has probably diminished in recent years, it is far from a thing of the past.

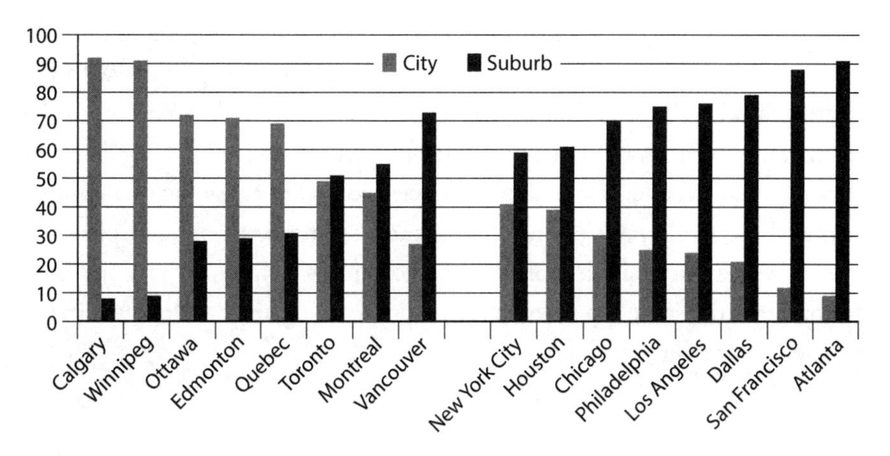

FIGURE 9-1. Urban/suburban population split in United States and Canadian metro areas
Sources: Compiled from Statistics Canada as adapted by Turcotte 2008; US Bureau of the Census.

Basic housing and other building types making up the suburban landscape are similar in Canada and the United States, with detached single-family houses making up the greater part of the suburban stock, but land use patterns in Canada tend to be more compact. Individual lots tend to be smaller and leapfrogging less common than in the United States. Newer outer suburbs in Canada, although typically developed at lower densities than their older, closer-in neighbors, are still built at much higher densities than their US counterparts. Typical lot widths in new suburban development in Canada tend to be 50 feet or less compared with 100 feet or more in many recent suburban developments in the eastern and midwestern United States.

Canadian suburban areas are also less dependent on cars. Although Canadians are highly car dependent compared with people in many European countries, Canadian metropolitan areas show higher rates of walking, bicycling, and transit use than their counterparts in the United States. Moreover, inner suburban communities in Canada tend largely to be medium- or high-value areas often situated inside the boundaries of the central city, in contrast to the pattern of often extreme contrasts between impoverished and affluent small, politically distinct suburban municipalities found in many inner suburban rings in the United States.

Although no single source can account for these differences, they ultimately appear to be driven by a series of closely related factors, beginning with the relationship between local and provincial government and extending to the manner in which, as a product of that relationship, local land use regulation is subordinated to larger regional or province-wide plans and policies and how, within those larger policies and plans, land use and transportation planning are integrated into a single, coherent whole. In this chapter, we explore those factors and offer some examples of suburban planning and development in the two countries to illustrate these points.

Governance, Authority, and Accountability

The fluidity of Canadian municipal boundaries reflects differences in the underlying authority structure under which local governments in Canada and the United States operate, differences that, although they may seem merely technical, are actually fundamental and have ramifications not only for the physical form, but the social and economic content of Canadian suburbs. Urban boundaries in Canada are far more likely to change through annexation or consolidation to reflect regional growth than in the United States; using David Rusk's terminology, they are far more "elastic"[2] than most US urban boundaries.

With rare exceptions, cities in the northeastern and midwestern states of the United States are locked into boundaries that were established in the 1920s or earlier, surrounded by separately incorporated suburban municipalities. The proliferation of such municipalities mystifies most observers. Bergen County, New Jersey, across the Hudson River from New York City, contains 74 separate incorporated municipalities and St. Louis County, Missouri, 90; Allegheny County, Pennsylvania, which contains Pittsburgh, and Cook County, Illinois, which contains Chicago, each contain more than 130 separate municipalities, each with its own control over planning, zoning, and land use.

Although the picture is somewhat different in the southern and western states, where cities are, at least in theory, permitted to annex surrounding areas as they grow or as utilities are extended, in practice it is often relatively easy for suburban areas to incorporate as separate municipalities, blocking annexation and locking in central-city boundaries. These states share the tendency of their eastern and midwestern counterparts to fragment their regions into growing numbers of small municipalities, although perhaps to a lesser degree. Maricopa County, Arizona, which contains Phoenix, has 26 separate cities, and Los Angeles County has 88. Arizona may be a land of wide-open spaces, but nearly four out of five of its residents live in incorporated cities.

Canadian metropolitan areas have far fewer separate municipalities. With a population of more than 6 million, the greater Toronto area has only 29 municipalities, more than any other region in Canada except for Montreal. Even that number is misleading, however, because under the Ontario system of upper- and lower-tier municipalities, comprehensive planning activities for most of those municipalities is carried out by four upper-tier regional municipalities.

In the United States, once a municipality has been incorporated, which is still fairly easy to do in many states, it is not likely ever to be dissolved. Although a few hundred municipalities, mostly ghost towns or postage-stamp communities with minute populations, have voluntarily dissolved over the years[3] and a smaller number have consolidated into larger city-county units like Nashville or Indianapolis, it is not clear that *any* municipality in the United States has ever been dissolved or merged with another by state government action.[4] Such is the status of the suburban municipality in the United States. Incorporation in the United States is effectively permanent and irrevocable short of a voluntary decision by the municipality's voters. Thus, incorporation can be seen, as Robert Wood wrote in his 1958 classic *Suburbia*, as a vehicle with which to erect "social and political barriers against invasion."[5] Although in a handful of cases states have taken over governance of municipalities that have fallen into fiscal disaster from which they have been unable to rescue themselves, as was the case with Detroit in 2013, these takeovers have always been short term, with municipal powers restored for better or for worse after a few years.

The picture is radically different in Canada. Rather than a right, incorporation can be seen as a privilege bestowed by the provincial government that can be withdrawn as easily as granted. Ontario was able to create the two-tiered municipality of Metropolitan Toronto in 1954 without a referendum and against the wishes of the municipalities involved and then completely merge the area into a unified City of Toronto in 1998 despite an overwhelmingly negative vote by the citizens of the affected municipalities. As a result of a 2001 consolidation, Ottawa today contains what had previously been ten distinct townships along with a handful of smaller cities such as Kanata, which had previously been incorporated as a city by carving out parts of three townships in 1978. All are now districts within the vastly larger city of Ottawa (figure 9-2).

The meaning of this difference goes beyond its practical implications. It leads to a fundamentally different sense of the relationship between the municipality and the province and to a different way of thinking about the exercise of municipal power. Simply stated, Canadian suburban officials, however much they may want to, cannot think of their city or village as being unrelated to the larger whole. Whatever autonomy it may possess is provisional and contingent, not permanent or inherent in its status as an incorporated municipality as is the case in the United States. Canadian municipalities share a culture of accountability.

The differences, however, are not only cultural. Canada maintains explicit mechanisms by which local government action is subordinated to or

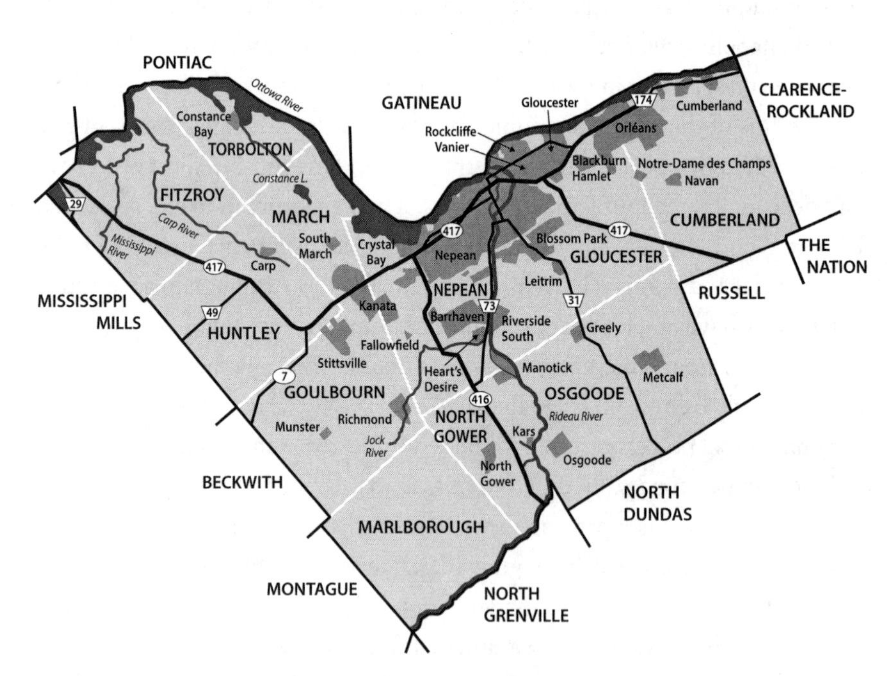

FIGURE 9-2. Map of Ottawa showing boundaries of former townships
Source: Wikimedia Commons, licensed under public domain.

constrained by policies adopted by higher levels of government. Although local governments have no fundamental constitutional status under either system, many states in the United States have granted local government considerable autonomy or home rule. Canadian municipalities, however, remain unequivocally creatures of provincial government; what Judith Garber and David Imbroscio wrote in 1996, that "the provincial hand remains firmly on the levers of land-use controls,"[6] remains true today. Canadian provinces exert a level of control over their municipalities' actions that would be unthinkable in any state in the United States.

The provinces have regularly exercised this control to create regional governance structures to manage growth and help finance strategic infrastructure. Ontario is a case in point. Provincial government maintains strong planning control over municipalities through provincial policy statements and, in the Greater Golden Horseshoe region surrounding Toronto, through regional planning conducted at the provincial level. Under the Ontario Planning Act, all local planning actions must comply with provincial growth management policies, which have been in place since the 1980s. The provincial Places to Grow plan, adopted in 2006, defines growth areas and strategic infrastructure plans and sets down density and intensification targets for the municipalities covered by the plan.[7]

Provincial planners work closely with upper-tier municipal planners to ensure that their plans faithfully reflect the growth management vision behind the provincial plan, reviewing and frequently requiring amendments to municipal planning documents to conform to provincial policies. Finally, parties objecting to local planning or land use decisions can appeal to the provincial Ontario Municipal Board,[8] an administrative body that can overrule municipal actions for many reasons, including inconsistency of the local bylaw or decision with provincial plans or policies.

Although every province has its own institutional idiosyncrasies, planning regimes in other Canadian jurisdictions are similar to that of Ontario. Most provinces have adopted policies governing major land use and transportation issues that municipalities must follow in their land use planning decisions, and a plethora of mechanisms for regional cooperation have been established, albeit with varying degrees of success, around almost every large Canadian city. Regional mechanisms can take the form of unicities under which the central city gradually annexes suburban and rural municipalities as the population grows, as in Winnipeg or Calgary; of two-tiered governance; or even of three-tiered governance, as in Montreal where lower- and upper-tier municipal governments participate in the Montreal Metropolitan Community (MMC) a provincially mandated metropolitan council responsible for strategic planning, and must conform local plans to the CMM regional development plan.

These frameworks reinforce the culture of accountability in Canada, where individual municipalities are accountable to varying degrees to regional and provincial bodies and must act in ways consistent with regional or provincial plans and policies. Although larger bodies are not *necessarily* more sensitive to larger growth management and environmental concerns,[9] they are more likely to pay attention to larger issues that may not be recognized at the local level. A framework in which infrastructure investment strategies are decided at the regional level and where local planning is accountable to regional or provincial-level bodies and must operate within parameters set by those bodies is significantly more likely to foster growth patterns that are sensitive to larger transportation, energy use, and environmental considerations.

Typical zoning densities in Canada are much higher than in most parts of the United States. Michael Lewyn, in his study of sprawl in the United States and Canada, comments disapprovingly of Hamilton's zoning ordinance that in its "'suburban residential zone', the minimum lot size is just over 5800 square feet, or roughly one-seventh of an acre," which he characterizes as an "anti-density regulation."[10] As we see later in this chapter, that would be considered extremely high-density zoning in much of the US. Table 9-1 illustrates the minimum lot size and frontage requirements in the zoning by-law of Niagara Falls, a city in Ontario of predominately suburban character with a population of roughly 83,000 and substantial amounts of undeveloped land in its boundaries. Single-family detached residential densities vary from roughly five to twelve units per acre, and multifamily densities go up to seventy-one units/acre.

The contrast with the predominant pattern in the United States could not be greater. A look at the town of Lewiston, New York, facing the northern part of Niagara Falls across the river is instructive. Lewiston's zoning ordinance contains no zoning districts that allow anything other than one- or two-family houses, and single-family houses are permitted on lots that range from 11,250

TABLE 9-1.
Zoning standards for residential development in Niagara Falls, Ontario

Zone	Permitted Uses	Minimum Lot Size	Minimum Frontage[a]
R-1	Single-family detached	3,440–8,600 ft^2 (five different R-1 zones)	33–59 ft
R-2	Single-family detached	3,970 ft^2	39 ft
	Semidetached (two-family)	3,220 ft^2/unit average 2,790 ft^2/unit minimum	28 ft/unit
R-3	Single-family detached	3,970 ft^2	39 ft
	Semidetached	3,220 ft^2/unit average 2,790 ft^2/unit minimum	28 ft/unit
	Townhouse	2,150 ft^2/unit	21 ft/unit
	Triplex (three-family)	2,650 ft^2/unit	69 ft/triplex
	Fourplex (four family)	2,530 ft^2/unit	82 ft/fourplex
R-4	Semidetached	3,230 ft^2/unit	30 ft/unit
	Townhouse	2,690 ft^2/unit	20 ft/unit (buildings with four or fewer units
	Stacked townhouse	2,150 $ft^{2\prime}$/unit	20 ft/unit (buildings with four or fewer units
R-5	Multifamily housing	Six different multifamily zones with maximum density varying from 20 units/acre to 71 units/acre.	

Source: Niagara Falls Draft Zoning By-Law Consolidation, January 2015.
[a] All values have been converted from metric measurements as they appear in the bylaw.

square feet to 25,000 square feet, depending on the zone and whether or not sewer service is available.[11] As we will see, even this zoning is modest compared with many suburban jurisdictions in the New York metropolitan area. These lot sizes are vast, however, by comparison to Canadian zoning; a not atypical suburban two-family duplex development near Toronto is shown in figure 9-3.

With communities having little or no accountability to higher levels of government or responsibility to address regional considerations, suburban land use regulation in the United States, as has been documented repeatedly beginning with Robert Wood and Norman Williams in the 1960s, is dominated by two parallel goals: preserving property values or real or perceived ways of life

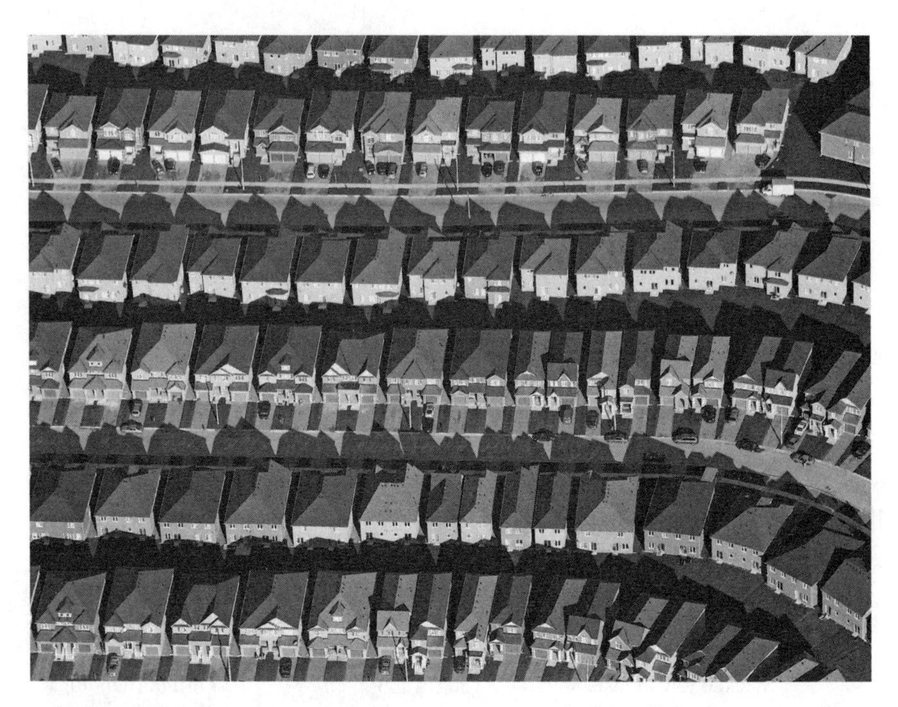

FIGURE 9-3. Suburban development in Markham, a suburb of Toronto
Source: IDuke on Wikimedia Commons Creative Commons Attribution-Share Alike 2.5 Generic License.

and competing for desirable land uses or property tax ratables. Both, albeit for different reasons, lead to exclusion on social or economic grounds. As a result, a major reason for the residents of an area to decide to incorporate is to be able to adopt their own comprehensive plan and zoning ordinance. The difference in the role of local zoning in Canada and the United States is not a function of a major difference in the mechanics of planning and zoning, but in the underlying policy and political framework in which it operates.

How suburban development takes place where higher levels of government lack effective tools to influence local practices is well illustrated by a 2011 study on suburban development trends in New Jersey.[12] New Jersey is particularly apropos. Not only is it the state with the highest overall population density in the United States and a distinctly suburban cast to its overall land use character, but since the enactment of the State Planning Act in 1986, it has had—at least on paper—a strong commitment to growth management embodied in a formal state development plan.[13] The State Development and Redevelopment Plan, among other matters, designates areas of the state in which development is to be encouraged ("smart growth" areas) and areas in which it is to be generally discouraged, while calling for development in the latter to be largely limited to incremental growth of existing population centers and compact, clustered development. Notably, however, the plan has avoided proposing explicit targets for intensification in smart growth areas, limiting itself to vague hortatory statements in support of that goal.

The 2011 study looked at development trends in the state of New Jersey followed by a closer look at Monmouth and Somerset Counties, two predominately suburban counties in the New York metropolitan area. It looked both at actual development patterns between 1986 and 2007 and at the zoning of the remaining undeveloped residential land.[14] The former analysis was done for the state and for the two counties, whereas the latter was done only in the two counties.

Figure 9-4 shows the strong trend toward very large lot development throughout New Jersey. Although prior to enactment of the State Planning Act only 24 percent of all residential land, including large amounts of urban land, had been developed at so-called rural densities of 1-plus acre lots (one house or

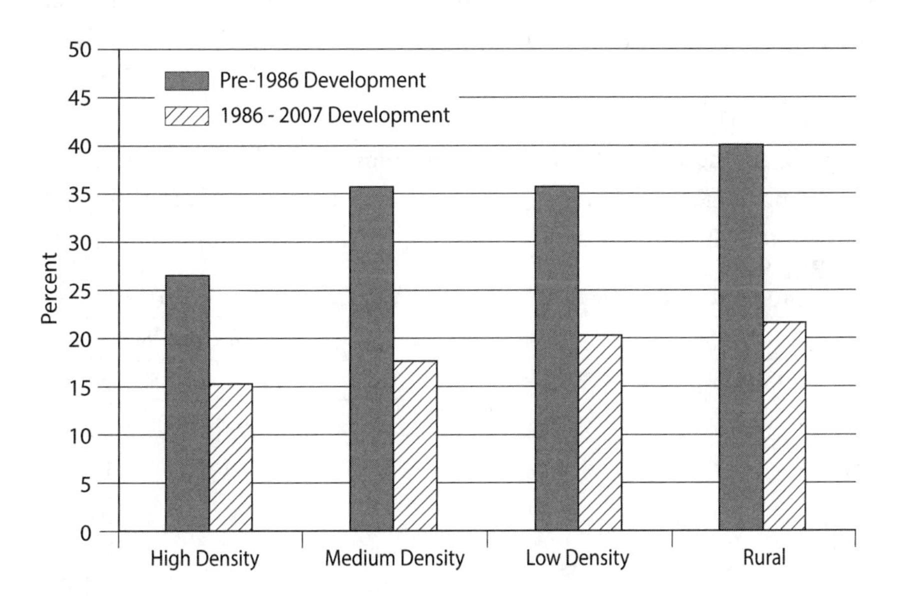

FIGURE 9-4. Residential development densities in New Jersey
Source: Compiled from Hasse, Reiser, and Picharz 2011.

fewer per acre),[15] 46 percent of additional land consumed by residential development between 1986 and 2007, or more than 100,000 acres, was developed for "rural" densities. This development is not, of course, rural by any reasonable definition; it is simply what has become the norm for suburban greenfields development in large parts of New Jersey and many other states, often in areas that could easily support higher density development.

The New Jersey study then calculated the zoning of the remaining undeveloped land in Monmouth and Somerset Counties, both of which still contain considerable undeveloped land. The prevalence of large-lot zoning in the remaining undeveloped land in these two counties, as shown in table 9-2, is even more striking. In Monmouth County, 84 percent of the land remaining for residential development is zoned for lots of 1 acre or larger, whereas less than 3 percent of the land is zoned for development at five units per acre or higher. In Somerset County, the picture is even more skewed: 87 percent of the remaining residential land is zoned for lots of 1 acre or larger, and only 1 percent is zoned for development at five units per acre or higher. Moreover, as the authors point out, most of this small amount of land was zoned in order to comply with the New Jersey Fair Housing Act, a statute that requires each municipality in the state to accommodate its "fair share" of housing for low- and moderate-income families. As the data show, this statute has only a negligible effect on the overall character of land use regulation.

In all, the nearly 67,000 acres of remaining undeveloped land in Monmouth and Somerset Counties is zoned to accommodate a total of only 48,000

TABLE 9-2.
Zoning of undeveloped land in Monmouth and Somerset Counties, New Jersey

	Monmouth County		Somerset County	
Density Category	**Acres**	**% of Total**	**Acres**	**% of Total**
High	1022	2.7	278	1.0
Medium	2936	7.8	950	3.3
Low	1920	5.1	2715	9.3
Rural	31583	84.3	25249	86.5
Total	37461	100%	29192	100%

Source: Hasse, Reiser, and Picharz 2011.

housing units, or one house for every 1.4 acres. If this land, all of which is well inside the perimeter of the New York metropolitan area and accessible to jobs and services, were zoned for a modest average of five units/acre, it would accommodate some 335,000 housing units.

On paper, New Jersey has a state plan and state policies that promote smart growth. In reality, not only has the lion's share of residential land consumption in New Jersey since adoption of the State Planning Act has been in areas designated in the state plan as inconsistent with smart growth, but almost all that development, as well as the zoning of the remaining undeveloped land, is for large-lot single-family development rather than for compact, clustered development as called for in the plan.

In other words, the local zoning and development pattern is not just inconsistent with the state plan and policies, but blatantly at odds with them. There are at least two distinct reasons for this. First, the state plan has no teeth; unlike Canadian municipalities, municipalities in New Jersey are not obligated to make their land use regulations fit the plan. The second reason lies in the counterproductive nature of the means by which the state chose to "enforce" its smart growth policies. Lacking the power to control local regulations or impose formal urban growth boundaries, the state adopted a policy of limiting the expansion of public sewer and water service areas and curtailing the use of state funds to expand roads and highways outside smart growth areas, two areas over which it has control. The outcome was predictable. In the absence of public sewer or water service or expanded highway networks, municipalities authorized only that development that could be accommodated with individual wells and septic tanks and served by minor roads, in other words, widely spaced, large-lot, single-family development. Needless to say, such development was completely in line with local preferences.

Integrated Planning

In previous chapters, we discussed many of the different approaches Canadian cities and metros use to make planning an integrated process, including the strong linkages between transportation and land use planning, and the use of development fees or exactions systematically as a planning tool. One might summarize the difference between the two countries as one in which Canadian communities see the regulation of land use as part of a larger planning system while those in the United States tend to see it more as an end in itself serving local interests. Clearly, as with every other difference we discuss, a continuum exists: although many suburbs in the United States do in fact integrate planning with land use to varying degrees, the tendency is in the other direction.

This result is not a function of the greater enlightenment of Canadian planners, but flows directly from the accountability structure described immediately above. Effective planning at the regional or metropolitan level and provincial oversight, coupled with the consolidation of municipalities into larger quasi-regional entities, as with Calgary or Ottawa, actively create the environment for integrated planning, something that is lacking in the great majority of US metropolitan areas.

Coordinated transit and land use planning in Canada goes back at least to the development of Toronto's Yonge Street subway line beginning in the 1950s, with the provision of density bonuses and air-rights leases around stations. Since then, many Canadian communities have used transit investment as a lever to guide urban growth, in particular to promote new regional town centers that will in turn support high levels of transit ridership, often done in advance of demand to further the goals of comprehensive regional plans. Rail lines have been extended in Toronto, Montreal, Calgary, and elsewhere explicitly to cluster growth around designated satellite subcenters, with rights-of-way reserved and protected in advance of construction in most cases and incentives such as upzoning and infrastructure improvements introduced simultaneously.

In Calgary, the city's unicity model, under which it gradually annexes suburban and rural areas as the city grows so that the entire developed area of the region falls within the city's boundaries, enables land use and transportation planning to be conducted under a single jurisdiction, indeed by the same municipal agency. Both land use and transportation plans are based on fostering nodal development around light rail stops, a goal gradually being realized thanks to the city's policy of spending half its transportation capital budget on transit system development.

Another policy difference that leads to better integrated planning and development in Canada than in the United States is the difference in the underlying legal frameworks governing developer exactions or impact fees. It is a critical issue because urban growth requires large-scale infrastructure investments: roads and bridges, water and sewer systems, schools, recreational and open space facilities, and much more. In both countries, some part of these costs are borne by land developers through mechanisms known as impact fees or exactions in the United States and development charges in Canada. The imposition of such fees serves the public interest in a number of ways, including making developers take responsibility for the external costs imposed on the public by their activities and by encouraging more efficient, compact development patterns, which result in fewer such costs.

With no need to address the problem of when a regulatory taking becomes unconstitutional or to establish the rational nexus, as it is known in United States law, between the public-sector costs generated by the development and the magnitude of the impact fee—neither are relevant under Canadian law—impact fees or development charges have become both significantly higher and significantly more consistently applied in Canada than in most US jurisdictions. Development charges tend to encompass many more types of infrastructure, including transit, schools, police and fire stations, waste management facilities, and even affordable housing and day-care facilities, and tend not to be discounted to the degree that is common in the United States.[16] Not surprisingly, then, typical charges in Canadian municipalities tend to be higher than in the United States, with changes between $40,000 and $60,000 per dwelling common for single-family developments in the Toronto suburbs.[17] In the United States, the national average for a single-family home in those jurisdictions with impact fees is $11,583, with California imposing the highest fees at an average of $22,154 per unit.[18] In addition to these fees, Canadian cities can require that the developer dedicate a percentage of the property as public open space, with provincial laws authorizing dedication of up to 5 percent of the property in Ontario and up to 10 percent in Quebec.

Many Canadian jurisdictions have adopted creative methods to calculate and apply development charges in ways that are less distorting of land markets and development patterns. For example, many suburban municipalities around

Toronto use "area-based" charges under which they distinguish between different parts of the municipality on the basis of the cost of providing infrastructure. Areas that are far from existing facilities or otherwise difficult to develop pay more in development charges than those in more infrastructure-efficient locations, with the charges based on land area consumed rather than on the number of dwelling unit rates. Ottawa assesses development charges on the basis of whether the development is within the greenbelt, outside the greenbelt, or in a rural area. To encourage transit-oriented development, the roads portion of the charge is reduced by 50 percent for projects near transit stops, and development charges in downtown areas are waived to encourage densification.[19] In Kelowna and several other British Columbia municipalities, development charges are assessed on the basis of the density of the project and its location within the municipality.[20] As a result, development charges play an important role in some Canadian jurisdictions in reinforcing growth management goals, something that is rare in the United States, where the charges are more likely to be seen strictly as a vehicle for raising revenue and where legal constraints may restrict municipal creativity.

In contrast to Canadian provinces, which provide explicit statutory guidance for development charges, guidance for US municipalities in setting such fees is more likely to come from court decisions than from clear state legal frameworks. Fees tend to be more limited than in Canada and vary from state to state, based on the widely varying standards adopted in state laws and court decisions. Under New Jersey law, for example, municipalities are allowed to levy impact fees only for the development's proportionate share based on the rational nexus test of sewer, water, road, and drainage costs. New Jersey falls in the middle ground in this area; in Illinois, the state supreme court required that an impact fee be "uniquely and specifically attributable" to the incremental need for infrastructure so that the exact users of the new capital facilities must be documented and a determination must be made of precisely how the fee paid relates to the need for capital facilities.[21] As one commentator writes: "It is extremely difficult to document the exact level of infrastructure need generated for each new development. When this test is used the courts nearly always strike down exactions."[22] By contrast, California courts have adopted a much more generous standard, known as the reasonable relationship test. Predictably enough, municipalities in California use exactions and impact fees far more extensively than in either New Jersey or Illinois and better integrate the different planning systems of land use, infrastructure, and transportation.

These variations all lead to significant limitations on the effectiveness of impact fees in the United States. To begin, many jurisdictions do not impose impact fees at all.[23] Second, those jurisdictions that do use impact fees generally limit the type of infrastructure that can be funded in this way to roads, water and sewer systems, and little else, not addressing the many other facilities that are needed to support growth, such as fire stations, libraries, waste management facilities, police stations, recreational buildings, and, of course, transit services.[24] Third, many municipalities further discount their fees, particularly on industrial and commercial development, well below the levels justified by objective cost analysis in response to pressure from developers or in the belief that fees will harm their competitive position for ratables vis-à-vis nearby jurisdictions. Abetting this situation is that many fees fall into areas of legal uncertainty because of the absence of clear statutory standards. Finally, charges are often based on a flat-rate or average cost approach, which fails to recognize that some developments are more efficient in terms of land use and infrastructure requirements than others. Thus, within a given municipality, high-density infill residential development in already largely developed areas may pay the same fees per unit as a large-lot detached house at the urban fringe, despite the obvious difference in new infrastructure requirements.[25]

As a result, impact fees as widely levied in the United States fail to accurately reflect the full costs of development generally or the variation in costs associated with different development patterns. Efficient users of land whose developments trigger lower per unit infrastructure costs subsidize inefficient users, providing no incentive for more efficient use of land and growth-oriented infrastructure and distorting urban housing and land markets. Ultimately, this system perpetuates more costly, in terms of both money and land consumption, development patterns.[26]

In contrast to the close coordination between transit and land use planning typical of regional growth planning in Canada, planning in the United States, with few exceptions, offers far less land use and transit coordination. That is not through lack of awareness of the importance of such coordination; the library of works calling for better coordination of land use and transportation in the United States would fill more than one bookcase. Although federal transportation funding law in the United States has led to the creation of a national network of regional agencies known as metropolitan planning organizations (MPOs) responsible for transportation planning, these agencies are largely toothless; as transportation guru Reed Ewing writes: "MPOs often do little more than rubber-stamp the transportation projects of state DOTs and local public works departments. Many take local land-use plans and projections as a given, to which they can only respond by building more roads."[27]

As local planning departments draft zoning ordinances and make development decisions, they are under no obligation to conform to regional plans or even to consult with the MPO or with transit service providers. As a result, the only transportation-related matters addressed in the local planning process may be the capacity of local roads to accommodate the car travel likely to be generated by the proposed development and the potential need to expand the road network to accommodate the demand. Because the demand is predicated on models that assume all but total car dependency, the outcome is predictable.

Although some coordination takes place in situations where fixed rail service already exists or is being developed, as in the Washington, DC, or Portland, Oregon, areas, little or no effort is made to integrate the expansion of transportation systems into the development of large areas in which such systems do not already exist. This pattern is exacerbated, on the one hand, by the stigma widely associated with conventional bus systems in the United States and, on the other hand, by the difficulty of obtaining financing for fixed rail systems or even for significant bus service expansion.

A case in point is the Route 1 corridor running through central New Jersey between Trenton and New Brunswick, a distance of roughly 20 miles. It is divided between two counties and seven municipalities. One of the counties falls under the Philadelphia metro MPO, and the other falls under the MPO serving the New York City area. The corridor is one of New Jersey's principal economic engines, containing nearly 10 million square feet of retail space and millions of square feet of office space, ranging from corporate headquarters to high-tech research complexes, almost all of which has been built since the 1970s.[28]

None of this development nor the underlying zoning that permitted it made any provision for transit, even though the linear configuration of the corridor would appear ideally suited for integrating land use and transportation planning. Although massive investments have been made to expand vehicle capacity on Route 1 itself, the only public transportation along the corridor is limited conventional bus service to a handful of major destinations.

As the state embarked on its honeymoon with smart growth in the 1990s, the Route 1 corridor became a matter of concern, leading to a 2003 decision by the state Department of Transportation (DOT) to initiate planning for a Bus Rapid Transit (BRT) route along the corridor. In the end, however, after years of planning and analysis and facing enormous technical, coordination, and financial obstacles, the project was sidelined, if not formally abandoned.[29]

The Route 1 story reflects many different issues. Each of the five principal municipalities along the corridor approved development largely without regard to the plans of other municipalities, and none accommodated transit in their plans. In the absence of any effective regional authority, no mechanism existed to compel municipalities to incorporate state or regional policies in local plans. Similarly, with no legal authority of their own, the two MPOs simply accepted the state DOT's policies—which for decades were limited to incremental expansion of the highway to accommodate more cars and trucks—as their own. By the time the DOT decided to explore transit solutions for the Route 1 corridor, it was effectively too late. With the existing highway right-of-way too narrow to accommodate BRT, the acquisition of a new right-of-way alone would have cost hundreds of millions of dollars.

Although it has many counterparts throughout the United States, the Route 1 corridor example is notable. First is the sheer magnitude of development that has taken place without consideration of transit options, and second is that eventually an effort was made to retrofit transit into the existing, largely developed, corridor, only to prove unsuccessful.

Even with changes in US policies and systems, so much development has taken place at densities far too low to support transit systems and so widely scattered across the region that change in many areas may be difficult, if not impossible. At the same time, however, significant opportunities exist not only in existing transit stations and hubs where additional development could be accommodated, but in the potential that will be created by the volume of future development likely to take place over the coming decades as older areas become ripe for redevelopment and reuse. In the absence of strong regional authority, however, the extent to which those opportunities are seized will depend largely, if not entirely, on the readiness of local governments to reflect regional thinking in their local plans. How likely that is will be seen in the next section, where we compare representative examples of suburban local planning in the United States and Canada.

Suburban Planning in the United States and Canada

It is difficult to form consistent comparisons between Canada and the United States with respect to suburban planning. With thousands of suburban municipalities in the United States and hundreds in Canada, of varying shapes, sizes, and degrees of development, almost any variety of planning approach is likely to be represented. At the risk of being accused of arbitrary selection, we have chosen to contrast the planning approaches followed in a small number of suburban jurisdictions in the two countries and provide a qualitative picture of the differences between the two, a picture that looks as much at how suburban municipalities *talk* about planning as about the *substance* of the plans. We look in various ways at four municipalities, two in Canada and two in the United States, as shown in table 9-3.

Few US suburbs are comparable in scale to a suburban city like Markham, north of Toronto, but one of the few is the town of Huntington, New York,[30] on suburban Long Island. Huntington is an affluent, sophisticated community, with a planning department led by an AICP[31]–certified planner. In 2008, the town adopted *Horizons 2020*, a comprehensive plan update produced with the assistance of a nationally respected planning firm.[32] The tone of the document

TABLE 9-3.
Municipalities discussed in this section

Municipality	Population	Density
Markham, Ontario	301,709	3,677/mi^2
Huntington, New York	203,262	2,162/mi^2
Brossard, Quebec	79,273	4,543/mi^2
Franklin, New Jersey	62,300	1,350/mi^2

is almost entirely inward-looking; of the four elements of the vision statement, as summarized in the plan's executive summary, two are directed at what can be best characterized as maintaining and enhancing the status quo:

1. Protect Huntington's small-town suburban character; preserve its rich heritage of historic resources; maintain and enhance its aesthetic character and identity, and practice responsible environmental stewardship.
2. Manage new development and redevelopment to protect neighborhood and village character, preserve open space, and revitalize commercial corridors; maintain a diverse employment base; develop an accessible, multimodal transportation system; and provide sustainable water, sewer, and stormwater infrastructure systems.

The other two elements of the vision statement are to provide a high quality of life and to provide responsive municipal government.[33] None is oriented toward change. The three guiding principles of the plan suggest a community that sees itself almost under siege:

1. Preserve those assets that exemplify Huntington's essential community character and quality of life, including the town's neighborhoods, villages, natural environment and remaining open spaces, history and heritage, arts, cultural life, and other assets.
2. Counteract trends that threaten community character and quality of life. Examples include escalating housing and school costs, visual blight along commercial corridors, limited variety of housing choices, negative environmental trends, and traffic congestion.
3. Implement strategies to maintain and enhance community assets and replace undesirable or obsolescent land uses with new ones that meet community needs, thus realizing the first and second principles.[34]

We are not suggesting that these principles are *wrong*, Instead, we simply believe that they look at the town of Huntington, a largely but not fully developed community built at almost entirely low suburban densities, from a defensive perspective, focusing on protection and preservation, rather than acknowledging the complex realities of the town's role as a large municipality in the heart of the principal economic region of the United States, located barely 15 miles from the boundary of New York City, today's preeminent global city, with frequent commuter rail service into Manhattan. Strategies that might reflect responses to this larger locational reality are largely absent from the more than two hundred pages of this document. Intensification, when it is mentioned, is not a strategy, but a threat to be countered.[35] Although Huntington is within the nominal jurisdiction of the Long Island Regional Planning Council, the council plays no role in the town's vision.

The significance of this planning approach becomes manifest when one looks at the current zoning of the town of Huntington (figure 9-5). Although there are scattered undeveloped areas in the town, there are no large expanses of land that have not been subdivided. With the exception of a large industrial area in the town's southwest corner, the town is made up almost entirely of residential development, with commercial development in a handful of older nodes and along arterial corridors. The town's housing stock is almost entirely single-family housing; 85 percent of all dwelling units were detached single-family homes in 2013, with almost all the rest either single-family townhouses or two-family homes.[36] The characterization of relatively extensive areas as "high-density" residential in the zoning ordinance is misleading in that almost all the land in those areas is in single-family development, albeit at modestly higher densities than in the rest of the town. No development other than single-family housing is permitted as an as-of-right use in this or any other zone.

The contrast with Markham, a suburban municipality of roughly similar size, situated in a similar relationship to Toronto, Canada's global city, as Huntington is to New York City. Markham's plan is as much about transformation as about protection, as the introduction to the city's 2013 comprehensive plan, *Planning Markham's Future*, makes clear:

> *During the postwar years, Markham, like most municipalities in the GTA [Greater Toronto Area], expanded rapidly through a series of ... developments that depended on the automobile for their success. In 1990, Markham embarked on a more sustainable model of development, going back to its historic village roots by planning for new compact, walkable communities ... combined with* intensification *along*

FIGURE 9-5. Generalized zoning map, Huntington, New York
Source: Town of Huntington, New York 2008.

major transit corridors and a new transit-based urban core Markham Centre, to accommodate additional growth. The focus of this Plan is to continue in this direction, with the goal of achieving a more urban, sustainable, complete City. (emphasis in original)[37]

To further this goal, Markham has identified a series of intensification areas, beginning with the creation of a walkable regional center in central Markham and including a cluster of key development areas and corridors (figure 9-6). To facilitate the intensification strategy, the city has adopted a plan for rapid transit improvements designed to ensure that the regional center, key development areas, and local centers are all served by rapid transit, with mobility hubs situated in strategic locations. Consistent with the regional approach to planning, the city's intensification strategy is designed to be consistent with targets established by the York Region's *2031 Intensification Strategy*, which in turn is designed to implement the overall policies set forth in the provincial Places to Grow plan. In designated centers, the Markham Official Plan calls for a minimum density of two hundred persons or jobs per hectare and a one-to-one resident-to-employee balance.[38]

FIGURE 9-6. Construction of higher-density housing in central Markham
Source: Raysonho on Wikimedia Commons, licensed under public domain.

Although most of Markham, from a land area standpoint, will continue to be in the residential low-rise category, Markham's plan calls for most future development to take place in the intensification areas, most significantly Markham Center, shown with the circle in figure 9-7. Notably, even in the lowest-density areas, semidetached, duplex, townhouse, and small multifamily (up to six-unit) developments are permitted, while the plan also calls for new development even in the low-rise areas to "be transit-oriented and reflect transit-oriented development principles."[39]

As figure 9-7 shows, a considerable part of Markham is still undeveloped. Although the majority of the undeveloped land is designated for future agricultural use or open space preservation, a large area has been designated as a future urban area, including a mix of residential and employment-oriented development. Future development of that area, which is proposed to be a mixed-use, walkable area with *minimum* residential densities of eight dwelling units per acre, is designed to take place, however, only when it is determined to be appropriate by York Region, in conformity with the provisions of the Provincial Growth Plan for the Greater Golden Horseshoe.[40]

One cannot tell from a planning document, in Markham or elsewhere in Canada, how local officials and planners feel about being part of a system in which they are accountable to regional and provincial authorities, in contrast to the relative autonomy of their US counterparts. The Markham Official Plan, however, certainly does not even contain an undercurrent of tension between the local and regional objectives and in one place even notes that "Markham's endorsed growth alternative to 2031 is more aggressive than the York Region Plan."[41] The plan projects the city to grow by 120,000 between 2011 and 2031, which is indeed an aggressive strategy, particularly one that depends on intensification rather than continued greenfields sprawl.

Similar contrasts can be found between the planning approaches in Brossard, an inner suburb of Montreal, and Franklin, New Jersey, a centrally located suburban township in the New York metropolitan area of similar population size. The contrast in tone between the plans adopted by the two municipalities is particularly notable: the 2006 master plan adopted by Franklin Township makes no pretense to be anything other than a technical document. That said, the township's plan is substantially more accommodating to varied land uses than that of Huntington, with two-family homes and low-rise multifamily housing or garden apartments permitted in a number of zones, albeit all largely developed. The plan also provides for redevelopment of the largely distressed area in the township's southeast corner, including opportunity for some limited intensification as a way to replace existing substandard and incompatible land uses.

In Franklin, although large-lot single-family development is permitted with minimum lot sizes ranging from 3 to 6 acres on the substantial amount of undeveloped land in the township, acquisition of development rights on agricultural lands and environmental constraints means that much of this land will remain undeveloped. As is the case in many US suburbs, this approach to farmland preservation results in more of a patchwork of farms and large-lot subdivisions than in a contiguous farming or open space belt.

In contrast to Huntington's all but solipsistic approach, Franklin's plan recognizes the existence both of adjacent municipalities and their plans as well as the State Development and Redevelopment Plan and addresses the goal of consistency, or at least avoiding patent incompatibility, with those plans. Franklin Township's master plan is a responsible attempt to address the ongoing needs of the community, but it is a narrowly technical document. It looks at the township from the inside rather than in the larger regional context, and it makes no effort to offer a larger planning strategy or a vision of what the community can or should become. In that respect, it falls short of the Huntington plan, which attempts to offer a vision of Huntington's future, however flawed and inadequate we consider that vision to be.

In contrast to Markham, which although clearly suburban in character is already a major center in the process of becoming a regional "edge city," Brossard is a more conventional suburb. Its main commercial artery, Boulevard Taschereau, is largely indistinguishable from a suburban strip in the United States, except the signage is in French rather than English. Most of the rest of the municipality is made up of neat subdivisions, interspersed with two- or three-story garden apartments, and a cluster of midrise apartment buildings along the St. Lawrence River.

Brossard, however, is in the midst of a massive planning process, designed, in effect, to reinvent the city. The process has been initiated with the 2013–2030 strategic plan, a bold, visionary document the tone of which is set by the slogan embodied on its cover "Urbaine/Contemporaine/Ouvert sur le Monde"

(Urban/Contemporary/Open to the World), to be followed by adoption of a new comprehensive plan and land use regulations designed to transform Brossard to a planned, intensified edge city, an "urbanized place on the periphery of a major urban center which assembles businesses, services, commercial centers, and offers a complete range of recreational and cultural activities."[42]

To that end, the plan sets a series of action elements, among which we find themes similar to those of Markham:

- Intensify principally in areas along public transit.
- Develop new areas and redevelop older areas, simultaneously taking into account mixed use, social diversity, transit accessibility, and proximity to services.[43]

Predictably, in light of the community's vision, the reconfiguration of Boulevard Taschereau is treated as a distinct goal and priority in itself[44] while the city is working with regional agencies to try to gain approval for a bus rapid transit line along the boulevard. The strategic plan, in addition to referencing the regional development plan of the MMC, notes that the regional plan "is and will be at the heart of the planning and development of our city over the coming years."[45]

Although Brossard has not yet adopted the official plan and land use regulations to implement the strategic plan, it has adopted a 2014–2017 action plan. It is also moving forward with its goal of adopting the new plan and regulations by 2016 and the action plan for Boulevard Taschereau by 2017.

Conclusion

Any comparison of the sort made in the preceding section can be accused of cherry-picking; indeed, there is no question that there are suburban municipalities in the United States doing enlightened, sophisticated planning that reflects an awareness and sensitivity to the larger needs and directions of the region and not just the desires of the municipality itself. In all likelihood, there are also Canadian suburbs that chafe under the strictures of provincial and regional plans and policies and, rather than embrace them, attempt to evade or work around them. Huntington's planning and development, one can argue, is narrowly focused even by comparison with other US suburbs; it is significant and relevant, however, because it is one of the very few US suburban municipalities that can be reasonably compared to Markham or other Canadian suburbs with respect to size and population as well as its similar spatial position within a comparable metropolitan area.

The point, though, is not that US suburban planning is exclusionary, narrowly focused, and indifferent to regional concerns and to such critical matters as the linkages between land use and transportation or that all Canadian suburban planning is the converse. Neither is true. Rather, the point is that the dynamics of the Canadian system foster suburban planning processes that lead to higher density and less sprawl, greater diversity of housing type (and affordability), and closer linkages between transportation, land use, and infrastructure.

Chapter 10
Learning from Canada: Conclusion and Recommendations

As we have described in this book, when it comes to fostering more compact, more energy-efficient and less auto-dependent suburban growth or stronger, more vital central cities and more sustainable urban regeneration, there is much that the United States can learn from the Canadian experience. Despite the considerable similarities in economies and political systems between the two nations, a combination of policies and practices, in part rooted in historical and social conditions, has led to markedly different outcomes for both suburban development and urban vitality. As we look at the reasons for those different outcomes, we recognize that not all of them are the result of factors that can be changed through specific policy recommendations. At the same time, however, many are. Before identifying areas where we believe that policy recommendations *are* appropriate and potentially feasible and laying out those recommendations, let us explore those distinctions and try to elucidate the policy climate within which proposed changes may or may not get a hearing.

The Framework for Policy Change

Our analysis of the variation in suburban and urban outcomes between Canada and the United States suggests three different ways in which that variation has come about. First are differences in *historic or social conditions*, second are differences in *past policy responses* at particular moments that may no longer be in place but that played an important role leading to present conditions, and third are *differences in current policies.*

An example of the first is the difference in the two nation's racial histories. Although it would be incorrect to say that race, or, more specifically, the problematic relationship between whites and African Americans in the United States, *drove* the course of urban decline and suburban expansion, racial strains and conflicts did have a significant role in shaping urban form in the United States in the years since World War II. Canada was fortunate to be far less affected by those strains, which as a result had far less effect on the course of Canadian development. More speculative, but still potentially significant, are arguments about differences in underlying national character or values, such as those discussed in chapter 2, stemming perhaps from differences in the manner in which the two nations were settled and won their independence.

Examples of the second are the far greater extent to which US cities engaged in urban renewal during the 1950s and 1960s and the radically different approaches to social housing pursued by Canada and the United States during the 1970s. Canadian cities escaped much of the devastation of urban renewal as well as much of the extreme poverty concentration that emerged in US cities, exacerbated by housing policies, during those years. Although cities in both countries still feel the effects of their nations' respective policies, the policies themselves—including, unfortunately, Canada's visionary social housing policies of that era—are long gone and unlikely to be resurrected. Similarly, a strong case can be made that the restrictive immigration policies in the United States, particularly during the 1950s and 1960s, either exacerbated or precluded an opportunity to reverse the decline of US cities. Although nothing can be done about the policies of that era, today, at a time when immigration reform is fitfully on the national agenda, it may be possible to craft new policies that might affirmatively further the revival of the still-lagging cities in the United States.

As we look at the differences in current policies, we must recognize that underlying political or legal differences may limit the extent to which the United States may be able to model changes in its policies to reflect the Canadian experience. Some issues, such as ones that emerge from the difference in the two countries' underlying constitutional frameworks, would not be practical to propose as recommendations. The takings clause in the Fifth Amendment to the United States Constitution may limit efforts to bring about a better approach to developer impact fees; as the California experience has shown, however, it may be a constraint, but not a barrier. Similarly, although the respective levels of crime and gun violence are a major difference in the quality of life in Canadian and United States cities, significant changes in gun control laws in the United States are well beyond the scope of a book that is grounded in a planning perspective, not to mention unlikely to happen in today's political climate.

The physical reality of large parts of the metropolitan landscape in much of the United States makes some of the changes that might be desirable in theory unrealistic in practice. Much suburban development in the United States, particularly that built since the 1970s, may simply be at too low a density, too scattered, and too dispersed across the region to be effectively served with public transit. Similarly, although we are convinced that much future urban revitalization could result from policy changes that may be attainable, we are realistic enough to acknowledge that other forms of change are likely to require levels of financial and political engagement from state and federal governments that may simply not be realistic in the foreseeable future.

Although the constraints are many, the picture is far from uniformly bleak. A notable feature of the scene in the United States today is how much things have changed in recent decades in ways that augur well for both more rational and sustainable suburban growth and for the revival of central cities. Urban revival in the United States is no longer limited to a few "hot" cities like Washington, DC, San Francisco, and Boston or scattered neighborhoods elsewhere. Once-depressed areas like the Warehouse District in Cleveland, Cincinnati's Over-the-Rhine, and Washington Avenue in St. Louis have become vibrant centers of activity. Pittsburgh has dramatically reclaimed its once-polluted riverfront and is converting its hillsides into a citywide greenway.

In suburban Virginia, coordinated land use and transportation planning linked to a public-private partnership is leading to a 1.3-million-square-foot mixed-use complex to be built in tandem with the opening of a new Metro station on the new line to Dulles International Airport. Edgewater, Florida, is moving forward with a new planned community that will incorporate a 3- to 4-mile streetcar line at a cost of only $10 million per mile, far less than the cost of retrofitting streetcar lines into existing development.[1] The suburban Denver municipality of Lakewood, historically a largely formless post–World War II suburb, created a downtown center by converting the declining Villa Italia suburban regional shopping mall into a vibrant mixed-use center, integrating retail, offices, and housing into a high-density, walkable environment.[2] Many more examples could be cited.

The total picture remains uneven, however. One city may be reviving in sound and sustainable fashion, but another may have little to show for hundreds of millions of dollars in public investment except for a stand-alone stadium, arena, or casino that creates few revitalization spin-offs or may even have a deadening effect on its surroundings. Although areas around some transit stations may bloom into diverse and walkable mixed-use centers, others remain fallow. Although some new suburban developments are clustered to provide walkable environments and offer potential for future transit service, scattered, large-lot development remains the rule rather than the exception at the suburban edge. The energy and creativity that exist to make cities and suburbs in the United States more livable and sustainable places, though, is palpable, and the achievements of practitioners and policy makers are real and significant in light of the limited tools and support with which they work. As they try to put into place smart strategies for future growth and redevelopment, they regularly face laws, policies, and bureaucratic obstacles that block their path. We are hopeful that our recommendations, growing out of the Canadian experience, will help spur changes that will strengthen their hands and foster a new generation of policies for livability and sustainability in US suburbs and cities.

What of the policy climate? Today, a reasonable observer could raise doubts about the extent to which meaningful policy change around sustainable planning and development is currently feasible in the United States. Washington finds itself in a state of ongoing conflict between a Republican Congress and a Democratic president, leading to deadlocks in many major policy areas, not least of which are renewable energy and climate change. At the grass roots, the Agenda 21 controversy, in which ideas such as smart growth, alternative energy, visioning, regional planning, and historic preservation have been demonized as being part of a United Nations plan for world domination,[3] is raging. Although seemingly far-fetched, the Agenda 21 controversy has been taken seriously enough by the American Planning Association for it to publish a bulletin for members of local planning commissions on how to address it.[4]

At the same time, there is another side to the story. Hundreds of thousands of highly educated and skilled young adults are moving back to the cities; while college-educated adults between the ages of twenty-five and thirty-four made up 4 percent of the population of the United States in 2013, they were 16 percent of the population of Washington DC, and 15 percent in Seattle and Boston. Their presence has transformed large parts of US cities and drawn massive new investment into mixed-used, transit-friendly areas. Historically car-dependent cities like Houston, Texas, and Tucson, Arizona, have built light rail lines as they try to turn their downtowns into higher-density, mixed-use places. Major universities from Connecticut to Arizona have invested millions in building "downtowns" on or adjacent to their sprawling campuses, such as the new Storrs Center adjacent to the University of Connecticut.

One could go on almost indefinitely. Los Angeles, once a byword for sprawl, is investing billions of dollars to create a truly urban transit system, while the Urban Land Institute, a prominent developer-oriented organization, has published a major report calling for rethinking infrastructure strategies in suburbia, pointing out that "as American suburbs build in more compact ways—with higher-density development clustered in nodes or along corridors and with increasing options for getting around without a car—reworking or rethinking infrastructure can be essential."[5] Consumer preferences are visibly shifting; although we are not among those who believe that the large-lot exurbs of the United States are the slums or ghost towns of tomorrow, we believe that evidence shows that neither are they the wave of the future.

The conclusion we reach after weighing the positive and negative features of today's political climate in the United States is that now is not a time to expect grand plans and major federal initiatives, but a time to work for gradual, incremental change, at the state level as much if not more than at the federal level. It is worth remembering that few if any of the Canadian practices we point to in earlier chapters are the result of federal initiatives; indeed, the federal government in office in Canada as we write has little more interest in the issues discussed in this book than does that of the United States. Moreover, we are not among those like Daniel Burnham who scorn small, incremental initiatives; we believe that there are many feasible steps that can be taken that can make a significant difference in both the short term and the long term. Some may be possible in some states and some in others, although there may be still other states where any meaningful change may be far distant. None of that is a bar to taking action where it is feasible to do so.

One final qualification is that our recommendations do not and are not meant to cover the full scope of what may be appropriate or desirable to foster livable and sustainable cities and suburbs in the United States. They are limited to those that flow from our analysis of the differences between the United States and Canada and where we can point to differences of policy and practice that have led to meaningfully different land use and planning outcomes. That two countries, similar in so many ways, could have such different outcomes is the lens through which these recommendations should be viewed.

Recommendations

In keeping with our conclusions about the constraints and opportunities offered by today's political climate, we propose a series of recommendations that we believe would significantly change the ground rules of development and redevelopment in the United States and foster greater livability and sustainability in three broad spheres:

1. Suburban greenfield development.
2. Suburban infill and intensification.
3. Urban revitalization and redevelopment.

Changes in all or any of these three areas could make a major different to the patterns of community growth in United States over the coming decades. Although infill and redevelopment may play a greater role in the overall development picture than in the recent past, it would be a serious mistake to assume that greenfield development, including continued suburban expansion in areas such as New Jersey's Monmouth and Somerset Counties, will not continue to play a major role in accommodating future population, household, and job growth. Even in the most urbanized states in the United States, large amounts of undeveloped land remain within commuting distance of urban areas and even closer to existing suburban or exurban job centers. Despite recent trends, we do not believe that the "return to the city," however real, will lead to the end of suburban greenfield development at any point in the foreseeable future.

One central theme pervades our recommendations. To foster compact and energy-efficient suburban growth while sustaining and enhancing central-city vitality, local planning and decision making need to be better integrated with higher levels of planning and policy making to reflect larger regional and statewide concerns. Doing so can take many forms, some of which may be acceptable in some jurisdictions and others elsewhere. To that end, our recommendations may be seen as a sort of menu of possibilities, all reflecting different dimensions of the Canadian experience.

Build a sustainable, policy-driven system for planning and regulating growth by integrating local planning into larger state and regional policy systems

Perhaps the single most powerful driver of compact, energy-efficient suburban growth and sustained urban vitality in Canada is the multilevel system by which planning takes place, grounded in the principle that local planning must fit into policies and strategies set at the regional or provincial level. Such systems are not unknown in the United States; there are a few areas, such as Portland, Oregon, or Minneapolis–St. Paul, that employ them to varying degrees, but these cities are outliers in an environment where local government decisions are rarely accountable to higher levels of government. Although the United States boasts a vast infrastructure of regional agencies—including metropolitan planning organizations, councils of government, and other acronymic bodies, many of which have adopted regional plans—and although a number of states have adopted statewide plans or growth strategies, few regional or state agencies have or have exercised the authority to ensure that municipal governments plan or act within their policy frameworks.

We have no illusions that state-level and regional planning bodies always act wisely and in the larger public interest; as are local entities, they are fallible and are subject to political and other influence. Still, the evidence from

Canada—and, we would suggest, from places where they have emerged in the United States—is that their presence leads to outcomes that are significantly better for the overall social, economic, and environmental health of their regions. To that end, states should either enact legislation to create regional planning agencies with the authority to adopt strategic regional development plans with which plans and planning decisions by municipalities within the region must be consistent, or they should give existing regional planning agencies those powers. Within that framework, regional plans should be required to follow the overall growth and development policies and standards established by the state government, including standards for establishing minimum densities of development for key areas served by adequate infrastructure and suitable for higher-density development than currently zoned; these standards would be embodied in regional plans and would be followed by municipalities in the development of their comprehensive plans.

State government policies and standards, however, should go beyond broad policy goals and address the following areas:

- Minimum target densities and other land use standards for key locations suitable for development, including areas close to transit and walkable centers and areas served by public sewer and water systems.
- Appropriate parking standards for different types of development under different conditions, including parking maximums.
- Integrating land use planning with transit and other alternatives to automobile use and with sewer and water infrastructure, including the adoption of complete street principles, both for new streets in greenfield development, and by retrofitting existing street systems.

To effectuate this system, state governments should designate regional agencies such as the Metropolitan Planning Organization for their respective regions and create administrative appeal bodies, similar in structure to the Ontario Municipal Board, the Washington Land Conservation and Development Commission, or the Massachusetts Housing Appeals Committee,[6] with the authority to rule on the consistency of municipal land use ordinance provisions and land use decisions with state policies and regional plans in a timely and efficient manner.

Create a strong legal and fiscal framework for integrated land use and transportation planning

The sine qua non for making future growth more transit-oriented and less auto-dependent is the integration of land use with transportation planning; considering how fundamental this principle is and how widely advocated it is by so many different people and organizations across the United States, it continues to amaze us how rarely it is put into practice. Only if future growth is linked to transit so that transit becomes a building block for compact, energy-efficient development, and density is increased in areas already well served by transit, will the United States be able to move significantly away from automobile dependency and energy wastefulness. Although there has been increased interest in fostering greater density around key nodes of existing transit systems in the United States through transit-oriented development and transit villages, systematic linkage of transportation and land use, even in existing systems, remains the exception rather than the rule; and when it happens, it typically reflects the preferences of local officials rather than larger strategies. Integrating land use and transportation planning requires that state or regional bodies have the authority to make sure that these issues are addressed in local plans; this requirement is even more important than many other areas of concern because transit systems outside large central cities typically cut across multiple municipal borders.

Integrating transit and land use, however, is not just about point-specific strategies such as creating high-density transit villages around transit stations, desirable as they are, but about integrating the transit *system* with the land use system to maximize ridership and foster larger patterns of transit-suitable development. In that respect, planning should focus as much on creating effective bus networks, linked to fixed-rail systems where they exist, rather than focusing largely or entirely on fixed-rail projects. Moreover, transportation planning should integrate measures to increase pedestrian and bicycle activity, particularly in higher-density areas where such measures are likely to be most effective; a good example is providing funds to create not only dedicated but separated bike lanes on commuting routes, as are common in Canada as well as in bicycle-friendly countries like Germany and the Netherlands. Such lanes would alleviate many of the safety concerns that hinder increased bike commuting.[7]

At the state level, within the regional planning framework discussed above, states should enact legislation to ensure the integration of land use and transportation planning, setting state policies to guide integrated planning and empowering regional agencies to adopt framework plans and standards that ensure that local plans and decisions are consistent with regional transit and land use plans. State laws should also provide that a significant share of the financial benefits to property owners from the extension or upgrading of transit systems is recaptured for public benefit, either to support transit or other important public activities associated with growth, such as supporting inclusion of affordable housing in development around transit stations.

At the federal level, federal transportation programs should provide incentives through increased transportation funding for projects in those states that have adopted appropriate laws and policies to integrate land use and transportation planning and in those regions that have implemented them. Major federal transit investments, either for construction of new transit lines, or for significant improvements such as new or extensively rebuilt stations to existing lines should only be available to states and regions where effective procedures for integrating land use and transportation planning are in place and being actively followed. We further recommend that federal highway funds should no longer be used for construction of new limited-access roadways, but only for maintenance and improvement of the existing highway infrastructure. If a particular city, county, or state sees a new highway as being critically important to its economic future, it should be willing to pay for it. We expect that these policies should and would lead to a significant redistribution of transportation funds away from highways and to transit systems.

Create stronger state machinery to further consolidation and reconfiguration of municipal boundaries and service delivery systems

Canadian provinces' readiness to use their authority to reconfigure municipal boundaries and redistribute responsibilities for delivering public services through two-tiered governmental systems is, in our judgment, one of the most important differences in governmental systems between the two countries and one in which the benefits are clearly apparent. That authority helps drive greater local accountability to larger planning and sustainability goals, reduce urban-suburban disparities, foster greater fiscal efficiency, and lead to more cost-effective and, in all likelihood, higher-quality public services.

Although state governments in the United States possess inherent powers to act in this area, we would be foolish not to recognize that the inviolability of municipal boundaries is something of a sacred cow in the United States. Although state governments have on occasion temporarily superseded local government powers in cases in which a municipality's fiscal or other incapacity put its citizens at risk, even these short-term interventions in extreme cases have tended to trigger protest.[8] Any exercise of state power to change municipal boundaries without local consent can be expected to be highly controversial.[9] At the same time, we believe that the benefits of doing so, both in terms of concrete changes as well as in terms of changing the accountability structure of local government, are manifest.

States should explore creating municipal boundary commissions that would recommend the consolidation of municipalities into regional entities as well as propose establishing two-tiered governmental entities, specifying the division of powers and responsibilities between lower-tier and higher-tier entities. Once the commission makes its recommendations, provisions should be built into state law to make it possible to implement the recommendations, even where affected municipalities may not support them. Such provisions might follow the model of the federal Base Closure and Realignment Commission (BRAC),[10] in which the commission is charged with recommending a package of military base closings or realignments based on statutory criteria, which are then submitted to Congress as a package. Congress can disapprove the package in its entirety, but cannot pick and choose from among the BRAC recommendations; if Congress does not disapprove it in forty-five days, the closures and realignments go into effect.[11] Similar procedures should be put into place with respect to school districts as well.

Municipal consolidation and reconfiguration are not simple or free of costs. Complicated issues inevitably arise— employee contract and pension provisions, responsibility for municipal debt, administrative procedures, and more—some of which cannot be resolved without cost, and the process of transition itself can impose costs. Any legislation designed to lead to consolidation of municipal boundaries and realignment of municipal functions should not only address how to harmonize disparities in the administrative, fiscal, and contractual structures of the municipalities involved, but also provide financial incentives to address those issues as well as the transition costs themselves. Incentives in themselves, however, without some form of mandate, have been shown to lead to no more than trivial changes in the dysfunctional boundaries and division of powers that characterize most parts of the United States.

Change state laws and policies governing municipal and school finances to make them more supportive of strong, healthy cities, and urban school districts

Many states in the United States provide different forms of assistance to municipalities and school districts; in some cases, state assistance is clearly designed to level the playing field between lower-income urban and more-affluent suburban jurisdictions. That is more widely the case with school aid, with many reluctant states being forced by state courts to provide state assistance, over their sometimes strenuous objections. The goal of the court decisions was, in most cases, to equalize total spending across rich and poor school districts. A good example is the Vermont Supreme Court decision that led to Act 60 in 1997, which created a state property tax pool and required contributions to the pool from wealthier municipalities, the state's so-called gold towns. Although Act 60 initially prompted strong opposition from those towns, it is part of the state's fiscal landscape today, and an independent evaluation done for the state legislature in 2012 found not only that was it a sound, equitable system, but that it had contributed to improvements in student performance in many parts of the state.[12] Elsewhere, though, state school aid is still based largely or entirely on population or pupil count.

Few states, however, provide as much assistance to school districts as Vermont does, and even more rarely do states provide the level of assistance that would enable central cities, as distinct from their school districts, saddled with both disparate service demands and limited fiscal resources, to provide high-quality public services and invest in high-quality infrastructure, both of which are critical conditions for urban vitality. Cities are not especially popular with most state legislatures, and few if any states have language in their state constitutions on behalf of cities analogous to the "education clause"[13] mandating the promotion of public education found in most state constitutions. The lack of adequate state support for urban municipal governments is compounded by other fiscal policies that limit the ability of local governments to raise revenues on their own and, in extreme cases, actually discriminate against urban areas in the allocation of costs between them and state government.[14]

A few decades ago, the argument that states should provide significant fiscal aid to urban municipal governments, if made at all, was made largely on the basis of social justice claims. Although those claims, which admittedly left many suburbanites cold, are no less valid today, an additional argument has become even more compelling. Changes in consumer preferences and the economy have given cities the potential to draw significant investment and economic growth and become the economic engines for their states' future growth; leveling the fiscal playing field is more likely to lead to that result than perpetrating continued fiscal disparities.

Where they do not already do so, states should provide financial assistance to school districts and municipal governments explicitly designed to equalize fiscal disparities between urban and suburban municipalities and school districts. Municipal aid as well as access for urban local governments to additional potential revenue sources should be designed to reward cities for better and more cost-effective performance with respect to their delivery of public services and their planning and implementation of strategies for revitalization.

Use impact fees as a rational, systematic approach to fund the public costs of development

It is widely, although perhaps not universally, accepted that it is appropriate to require developers to contribute to mitigating public costs associated with their development activities; it is also generally agreed that the imposition of those costs on developers should be consistent and reasonable. The fundamental test for such costs was set down by the United States Supreme

Court in its 1994 decision in *Dolan v. City of Tigard* in which it held that "no precise mathematical calculation is required, but the city must make some sort of individualized determination that the required dedication is related both in nature and extent to the impact of the proposed development."[15] The court further noted that its decision was consistent with the "reasonable relationship" standard applied by most state courts.

Public costs associated with development are not limited to the provision of roads, sewers, or public water supply to the development. They include the incremental cost of adding school children to the school system as well as parkland and recreational facilities to the open space network. They can include the marginal costs of increasing operational capacity to address the growth taking place. All of them, set at reasonable levels on the basis of rational standards and applied consistently as developments seek approval, should be considered legitimate subjects for impact fees or, as they are often known, exactions. Exactions not only mitigate municipal cost burdens; they may also act as incentives to prompt developers to pursue more compact and higher-density development.

California arguably has a system of exactions that approximates that standard. The definition of legitimate impact fees is broad but not unreasonably so, and their imposition is governed by state law. Although state courts have ruled on impact fees often, upholding some and overturning others, the underlying state statutes give municipalities considerable clarity about what is and is not acceptable, as distinct from some other states where making that determination is not unlike reading tea leaves. In other states, statutes define the realm of permissible exactions far more narrowly; an example is New Jersey, where the only areas in which a municipality is permitted to require fees are sewer, water, roads, and drainage. This restriction is not because of any state constitutional limitation, but purely because of the predilections of the state legislature or the influence of certain interests on their actions.

State legislation in this area is critical because of the need to promote consistency and transparency and to avoid unpredictable, case-by-case local negotiations. States should enact legislation or, where necessary, amend existing statutes to permit local governments to impose comprehensive impact fees that accurately and adequately address the public costs associated with growth and development. They should also set forth consistent and predictable procedures by which both municipalities and developers can determine the appropriate level of costs or other benefits, such as parkland dedication, to be paid in each case. In addition, state laws should provide incentives for development fostering greater sustainability and clear social benefits, such as fee waivers for affordable housing and reduced fees for infill developments and higher-density, mixed-use development projects.

Change tax policies to reduce incentives for sprawl and dependence on automobiles

Taxation is a powerful driver of policy, and tax policies have a powerful effect on development decisions and vehicle use. Two tax policies in the United States are particularly important in that respect. First, low state and federal gasoline taxes encourage automobile use and render public transit options less competitive than auto travel. Gasoline taxes in the United States, although varying widely from state to state, are roughly only one-third to one-half of those in Canada, which in turn are lower than typical gasoline taxes in European countries. Second, the federal income tax deduction on home-mortgage interest contributes to sprawl by encouraging people to buy larger and more expensive homes than they might otherwise buy. This tax deduction is highly regressive in that it creates upward pressure on prices throughout the market. Although lower-income home owners rarely take the deduction, they are nonetheless forced to pay more for housing as a result without gaining the offsetting benefits of the deduction. In addition, although it pushes up house prices, it has no visible effect on home ownership rates, and widespread representations to that effect, mostly coming from real estate agents and developers, have no foundation and are no more than special-interest pleading. Home ownership rates in Canada as well as in many European countries, where no such deduction exists, are comparable to or higher than those in the United States. Both low gas taxes and home-mortgage interest tax deductions deprive the public sector of valuable revenues while offering no offsetting public benefit.

The federal gasoline tax has been steady at 18.4 cents per gallon from 1993 to the present, during which period it has lost nearly 40 percent of its purchasing power. The federal government should increase that tax, as should those states such as New Jersey, Missouri, and South Carolina, all of which have state gasoline taxes below 20 cents per gallon. In parallel with these increases, both individual and employer incentives for use of public transit should be expanded. The federal government should also abolish the federal income tax deductibility of home-mortgage interest. Because large numbers of today's home owners bought their homes on the assumption that the deduction would be available and would experience financial hardship if it were suddenly to disappear, some form of gradual phasing out of the deduction for existing home owners should be provided.

Increase opportunities for mixed-income and mixed-use development

Attitudes toward both mixed-income and mixed-use development in the United States have changed dramatically since the 1980s. Although not long ago developers, planners, local officials, and others were extolling the benefits—even the necessity—of maintaining strict segregation in development by use and by income range, the desirability of mixed-income and mixed-use development has become widely recognized and may soon be considered the conventional wisdom. There is good reason for this change. Fostering social equity and opportunity, basic principles of sustainable development, demand that, to the extent feasible, future residential development be mixed-income development in which low- and moderate-income households benefit from proximity to transit, good public services, and job opportunities while growing communities can benefit from a diverse workforce. Mixed-use development is not only less dependent on cars and less conducive to sprawl, but also reflects growing consumer preferences. Indeed, it can reasonably be argued that mixed-use development represents the normal course of development in response to human needs and desires, whether in the United States or elsewhere, and that the fetishizing of separated land uses in the United States since the mid-twentieth century is the aberration and is now beginning to right itself.

As attitudes have undergone a sea change, however, actual practices have not. Although planning and development publications point to examples of mixed-income and mixed-use development, they are still the exceptions to the rule, and such development is still widely hindered by both legal and fiscal obstacles. In addition, although inclusionary zoning is widely used in some parts of the United States, it is used far less than it could be; in fact, many states still place obstacles in its path, as when state laws banning rent control in the private market have been interpreted by the courts as barring inclusionary rental housing. The rules that govern the only large-scale federal subsidies for production of affordable housing, the Low Income Housing Tax Credit (LIHTC), make it extremely difficult to use those subsidies either to create mixed-income developments or to incorporate LIHTC units into larger mixed-income developments. As a result, despite increasing recognition of the value of economic diversity, the overwhelming majority of LIHTC developments being constructed, whether in urban or suburban areas, continue to be discrete, separate, and means-tested low-income developments. Similarly, development of mixed-income projects and conversion of existing single-use buildings to mixed use continue to be hindered, particularly outside large central cities, by single-use zoning and land use regulation as well as by financing constraints. Those constraints reflect both federal regulatory obstacles as well as the shortage of well-established vehicles for mixed-use development financing in the private sector.

Changing practices to reflect changing attitudes will require both federal and state action. At the federal level, the LIHTC should be restructured to make it easier to use the program both to create mixed-income developments and to facilitate inclusion of low-income rental units in larger mixed-income projects. Although changing the LIHTC will require legislation, other changes can take place through administrative action. Regulations governing Federal Housing Administration mortgage insurance and the purchase of mortgages by Fannie Mae and Freddie Mac should be revised to recognize that mixed-use development is both desirable and normal and to eliminate arbitrary constraints on the percentages of nonresidential development in mixed-use developments where developers are seeking mortgage insurance or access to the secondary market.

State governments should remove restrictions on inclusionary development, such as rent-control prohibitions that have been interpreted as barring inclusionary development, and in those states where enabling legislation is required for municipalities to enact such ordinances, the state government should enact explicit language authorizing the use of inclusionary provisions in local zoning ordinances. State housing finance agencies, which have the responsibility under federal law of allocating the federal tax credits provided under the LIHTC program, should give preference to mixed-income projects as they allocate their credits.[16]

Finally, in tandem with changes from federal agencies, it is essential for the US lending industry to reconsider its approach to financing mixed-use development. New urbanist John Norquist has characterized the situation well. He writes that after World War II, his uncle

> *bought a two-story building on Payne Avenue on the east side of St. Paul. It had two apartments on the second floor and space on the first floor for his plumbing company. When he applied for a loan, the banker was pleased that the building included two apartments. It reduced risk in that if the plumbing business took a while to become profitable, the apartments would provide cash flow in the meantime. But if my uncle tried to borrow money for his building today, he would likely hear a different message from his banker. The bank would question the viability of a building that contained both a business and housing, as one or the other might fail and diminish the prospects for a return on capital and repayment of the bank's loan. Instead of looking at the diversity of uses as a way to reduce risk, nowadays mixing of uses is considered high-risk.*[17]

The number of mixed-use developments that are somehow successfully cobbled together by developers despite these constraints testifies to the extent to which the consumer market is looking for this product, which in turn represents a critical element in making sustainable development a reality in US towns and cities. The financial industry should treat investment in mixed-use development as a clearly defined asset class, taking advantage of the solid information base that already exists about the risks and rewards of this type of development.

Build measures to foster immigrant investment and settlement in distressed urban areas into future immigration reforms

Immigration reform may appear somewhat distant from the focus on land use, infrastructure, and transit that has permeated this book, but it is particularly germane, not only because of the significant differences between the United States and Canada and their implications for the health of urban centers, but also because the prospects for federal reform, in some fashion, of immigration law in the United States appear greater than those for major legislation in almost any other national domestic policy issue. Although there continues to be heated disagreement over the contours of reform, here there is, in contrast to other issues, something close to a consensus that the system is broken and that reform is needed.

Immigrants have made an immeasurable contribution to Canadian cities since World War II. As we discussed earlier, the significantly higher level of immigration to Canada from the 1950s through the 1970s compared with the United States and the extent to which those immigrants settled in central cities bear greatly on the vitality of Canadian cities today, and those trends are continuing. As of 2001, nearly 60 percent of all Canadian immigrants, and 70 percent of those arriving in the preceding decade, lived in Toronto, Vancouver, and Montreal. Although one cannot change the past, one can try to influence the future. Greater immigration to distressed older cities in coming decades could significantly improve prospects for their regeneration. Many US cities, including Detroit and Philadelphia, have come to recognize this fact and are actively pursuing efforts to become more attractive immigrant destinations.

As immigration reform continues to be a subject for discussion, the urban opportunity that it represents needs to be made part of the conversation. Incentives to encourage immigrants to settle in distressed urban centers, particularly immigrants who bring specialized skills and resources and to increase immigrant investment in those centers, perhaps built on existing but narrowly limited programs that exist in current law, should be included in any comprehensive federal legislation enacted in this area.

Closing Note

This book has shown that policy differences between the United States and Canada—two countries that are strikingly similar in many respects—have had powerful effects on urban and suburban growth and development and on the forms that they have taken. Cities and suburbs in Canada and United States are much more like one another than their counterparts elsewhere in the world, but they nonetheless differ in important ways. Although the differences are rooted in part in cultural and social differences between the two societies, they are as much the product of policy differences. These differences have important implications and offer valuable lessons for those concerned with promoting the future vitality and sustainability of cities and suburbs in the United States.

The United States is a large, diverse, and prosperous country with a rich history of planning and development, and there are many features of its growth and development in which planners, public officials, and developers can take pride. Indeed, as we regularly note, many of the Canadian practices that we cite with approval are already being pursed in at least some jurisdictions in the United States. The distinction, however, is that practices that are the norm, or close to it, in Canada are outliers in the United States, just as by many measures US cities and metropolitan areas are outliers compared with their counterparts in other developed nations.

The United States has an opportunity to learn from Canada, an opportunity that is made greater by the considerable similarity between the legal and economic systems of the United States and Canada in contrast to the far greater differences between the systems of the United States and those of most European nations. The United States can learn much from Canada with respect to the legal and fiscal structures and public policies that foster more sustainable suburban development, stronger transit systems better linked to growth and development, reduced car dependency, thriving urban centers, and lively, compact suburbs. Our recommendations flow from our assessment of what those structures and policies are and how their essence might be successfully fostered within the political and legal systems of the United States and its many states.

Change in development and settlement patterns does not come quickly or easily. We believe, though, that the steps we call for, based on the Canadian experience, would contribute greatly to putting the United States on a path toward meaningful change in urban form and to important social, economic, and environmental benefits that would flow from that change. Change is already taking place in many parts of the country, although more through guerilla action by developers, advocates, and scattered state and local governments and through a new generation of citizens voting for the cities with their feet than through coherent state or federal policy.

It is time to see policy catch up with the movement on the ground. Although many of our proposals are challenging, and we are certainly aware of the serious obstacles standing in the way of their adoption, we are also aware that they are consistent with and can only reinforce many of the strong, positive emerging trends visible in cities and metros across the United States. For that reason, we close by reiterating our optimism that ideas and proposals such as those we advocate can and will be more broadly applied over the coming years. In short, the United States might gain a great deal by looking to its north.

Notes

INTRODUCTION

1. David M. Thomas and David N. Biette, eds., *Canada and the United States: Differences That Count*, 4th ed. (Toronto: University of Toronto Press, 2014).
2. Michael Adams, *Fire and Ice: The United States, Canada and the Myth of Converging Values* (Toronto: Penguin Canada, 2003).
3. Seymour Martin Lipset, *Continental Divide: The Values and Institutions of the United States and Canada* (New York: Routledge, 1990).
4. Although this impression has not been verified statistically, it is our impression that a disproportionate share of the comparative research has been written by scholars raised and trained in the United States; they subsequently relocated to academic positions at Canadian universities.
5. Maurice Yeates and Barry Garner, *The North American City* (New York: Longman, 1976).
6. Michael A. Goldberg and John Mercer, *The Myth of the North American City: Continentalism Challenged* (Vancouver: University of British Columbia Press, 1986).
7. Goldberg and Mercer, *The Myth of the North American* City, 239.
8. Frances Frisken, "Canadian Cities and the American Example: A Prologue to Urban Policy Analysis," *Canadian Public Administration* 29, no. 3 (1986): 345–76.
9. John Mercer and Kim England, "Canadian Cities in Continental Context: Global and Continental Perspectives on Canadian Urban Development," in *Canadian Cities in Transition,* 2nd ed., eds. Trudi Bunting and Pierre Filion (Don Mills, ON: Oxford University Press, 2000), 55–75.
10. Zachary Taylor, "The Politics of Metropolitan Development: Institutions, Interests, and Ideas in the Making of Urban Governance in the United States and Canada, 1800–2000" (PhD diss., Department of Political Science, University of Toronto, 2015).

11. John Miron, "Urban Sprawl in Canada and America: Just How Dissimilar?," paper presented at the annual meeting of the Association of American Geographers, New Orleans, LA, March 2003, http://www.utsc.utoronto.ca/~miron/Miron2003USCAJHD.pdf; and Michael Lewyn, "Sprawl in Canada and the United States," *Urban Lawyer* 44, no. 1 (2012): 85–133.
12. John Pucher and Ralph Buehler, "Why Canadians Cycle More Than Americans: A Comparative Analysis of Bicycling Trends and Policies," *Transport Policy* 13, no. 3 (2006): 265–79.
13. J. R. Pucher, "Public Transport Developments: Canada vs. the United States," *Transportation Quarterly* 48, no. 1 (1994); and Robert Cervero, "Urban Transit in Canada: Integration and Innovation at Its Best," *Transportation Quarterly* 40, no. 3 (1986): 293–316.

CHAPTER 1

1. We will use the term *cities* to refer to both central cities and urban regions, including central cities and surrounding suburbs. To avoid confusion between these two common meanings of the word, we make an effort to clarify which meaning we intend in each context where it is used.
2. When referring to the United States and Canada together, we've chosen the term *Northern American* as opposed to *North American* to avoid the implication that we are including Mexico and the dozens of other countries and overseas territories that make up the whole continent.
3. Arthur Nelson, *Reshaping Metropolitan America: Development Trends and Opportunities to 2030* (Washington, DC: Island Press, 2013).
4. Nelson, *Reshaping Metropolitan America*.
5. Nathan Norris, "Why Gen Y Is Causing the Great Migration of the 21st Century," *PlaceShakers and Newsmakers*, April 9, 2012.
6. Nelson, *Reshaping Metropolitan America*, 81.
7. Arthur Nelson, "Resetting the Demand for Multifamily Housing: Demographic and Economic Drivers to 2020," Presentation to National Multi Housing Council, 2010.
8. See, e.g., Robert Bruegmann, *Sprawl: A Compact History* (Chicago: University of Chicago Press, 2005).
9. Robert Margo, "Explaining the Postwar Suburbanization of the Population in the United States: The Role of Income," *Journal of Urban Economics* 31 (1992): 301–10.
10. Pew Research Center, "The Lost Decade of the Middle Class," *Social and Demographic Trends*, August 22, 2012.

11. Edward N. Wolff, *Household Wealth Trends in the United States, 1962–2013: What Happened over the Great Recession?* NBER Working Paper No. 20733 (Cambridge, MA: National Bureau of Economic Research, 2014).
12. Marina Vornovytskyy, Alfred Gottschalck, and Adam Smith, *Household Debt in the U.S., 2000 to 2011* (Washington, DC: US Census Bureau, n.d.).
13. Anthony Leiserowitz, Edward Maibach, Connie Roser-Renouf, and Peter Howe, *Extreme Weather and Climate Change in the American Mind* (New Haven, CT: Yale Project on Climate Change Communication, 2012).
14. Edward Maibach, Connie Roser-Renouf, Emily Vraga, Brittany Bloodhart, Ashley Anderson, Neil Stenhouse, and Anthony Leiser, *A National Survey of Republicans and Republican-Leaning Independents on Energy and Climate Change* (New Haven, CT: Yale Project on Climate Change Communication, 2013).
15. Giles Parkinson, "HSBC: World Is Hurtling towards Peak Planet," *RenewEconomy*, March 27, 2013.
16. Secretary of Defense, *Quadrennial Defense Review* (Washington, DC: US Department of Defense, 2014).
17. The US emissions are 17.2 metric tons per capita in 2009, compared with 5.3 for Sweden, 9.6 for Germany, 8.5 for the United Kingdom, and for 6.3 France.
18. Pietro Nicola, *Laws of the Landscape: How Policies Shape Cities in Europe and America* (Washington, DC: Brookings Institution Press, 1999).
19. David Owen, *Green Metropolis: Why Living Smaller, Living Closer, and Driving Less Are the Keys to Sustainability (New York: Riverhead Books, 2010).*
20. D. Hernandez, M. Lister, and C. Suarez, *Location Efficiency and Housing Type: Boiling it Down to BTUs* (Washington, DC: US Environmental Protection Agency, 2011).
21. Cambridge Systematics, Inc., *Moving Cooler: An Analysis of Transportation Strategies for Reducing Greenhouse Gas Emissions (Washington, DC: Urban Land Institute, 2009).*
22. Jeff Speck, *Walkable City: How Downtown Can Save America, One Step at a Time* (New York: Farrar, Straus and Giroux, 2012).
23. Richard Florida, *The Rise of the Creative Class* (St. Louis, MO: Turtleback Books, 2004).
24. Jamie Peck, "Struggling with the Creative Class," *International Journal of Urban and Regional Research* 29, no. 4 (2005): 740–70.
25. American Society of Civil Engineers, *2013 Report Card for America's Infrastructure* (Reston, VA: American Society of Civil Engineers, 2013).
26. Robert Burchell et. al., *The Costs of Sprawl—Revisited,* prepared for the Transportation Research Board of the National Research Council (Washington, DC: National Academies Press, 1998).

27. Robert Burchell, G. Lowenstein, W. R. Dolphin, C. C. Galley, A. Downs, S. Seskin, K. G. Still, P. Brinckerhoff, and T. Moore, *Costs of Sprawl—2000* (Washington, DC: Transportation Research Board, 2002).
28. Reid Ewing and Robert Cervero, "Travel and the Built Environment—Synthesis," *Transportation Research Record* 1780 (2001): 87–112.
29. Andrew Dannenberg, Howard Frumkin, and Richard Jackson, *Making Healthy Places: Designing and Building for Health, Well-Being, and Sustainability* (Washington, DC: Island Press, 2011): xvii.
30. Lawrence Frank and Peter Engelke, "Multiple Impacts of the Built Environment on Public Health: Walkable Places and the Exposure to Air Pollution," *International Regional Science Review* 28 (2005): 193–216.
31. Lawrence Frank, Brian Stone, and William Bachman, "Linking Land Use with Household Vehicle Emissions in the Central Puget Sound: Methodological Framework and Findings," *Transportation Research Part D* 5, no. 3 (2000): 173–96.
32. A. Goonetilleke, E. Thomas, S. Ginn, and D. Gilbert, "Understanding the Role of Land Use in Urban Stormwater Quality Management," *Journal of Environmental Management* 74 (2005): 31–42.
33. R. Ewing, R. Schieber, and C. Zegeer, "Urban Sprawl as a Risk Factor in Motor Vehicle Occupant and Pedestrian Fatalities," *American Journal of Public Health* 93, no. 9 (2003): 1541–45.
34. Margie Peden, Richard Scurfield, David Sleet, Dinesh Mohan, Adnan A. Hyder, Eva Jarawan, and Colin Mathers, eds., *World Report on Road Traffic Injury Prevention: Special Report for World Health Day on Road Safety* (Geneva: World Health Organization, 2004).
35. L. Champion. "The Relationship between Social Vulnerability and the Occurrence of Severely Threatening Life Events," *Psychological Medicine* 20, no. 1 (1990): 157–61.
36. Russ Lopez, "Urban Sprawl and Risk for Being Overweight or Obese," *American Journal of Public Health* 94, no. 9 (2004): 1574–79.
37. R. Sturm and D. Cohen, "Suburban Sprawl and Physical and Mental Health," *Public Health, Journal of the Royal Institute of Public Health* 118, no. 7 (2004): 488–96.
38. Keith Lawton, "The Urban Structure and Personal Travel: An Analysis of Portland, Oregon, Data and Some National and International Data," paper presented at the E-Vision 2000 Conference, Washington, DC, October 11–13, 2001.
39. Andrew Dannenberg, Howard Frumkin, Richard Jackson, eds., *Making Healthy Places: Designing and Building for Health, Well-Being, and Sustainability* (Washington, DC: Island Press, 2011).

40. Chris Benner and Manuel Pastor, "Brother, Can You Spare Some Time? Sustaining Prosperity and Social Inclusion in America's Metropolitan Regions," *Urban Studies* 52, no. 7 (2015): 1339–56.
41. Thad Williamson, *Sprawl, Justice, and Citizenship: The Civic Costs of the American Way of Life* (New York: Oxford University Press, 2010).
42. Elizabeth Kneebone and Alan Berube, *Confronting Suburban Poverty in America* (Washington, DC: Brookings Institution Press, 2013).
43. Carmen Arroyo, *The Funding Gap* (Washington, DC: Education Trust, 2008).
44. Joseph Persky and Wim Wiewel, *When Corporations Leave Town* (Detroit: Wayne State University Press, 2000).
45. Michael Stoll, "Spatial Mismatch and Job Sprawl," in *The Black Metropolis in the Twenty-First Century: Race, Power, and Politics of Place*, ed. R. D. Bullard (Lanham, MD: Rowman and Littlefield, 2007), 127–48.
46. See, for example, Rolf Pendall and Christopher R. Hayes, *Driving to Opportunity: Understanding the Links among Transportation Access, Residential Outcomes, And Economic Opportunity for Housing Voucher Recipient* (Washington, DC: Urban Institute, 2014).
47. Reid Ewing and Shima Hamidi, *Measuring Urban Sprawl and Validating Sprawl Measures* (Washington, DC: National Institutes of Health and Smart Growth America, 2014).
48. Nicola, *Laws of the Landscape.*
49. Bruce Katz and Joel Rogers, "The Next Urban Agenda," in *The Next Agenda: Blueprint for a New Progressive Movement,* eds. R. Borosage and R. Hickey (Boulder, CO: Westview, 2001), 189–210.
50. Roberta Brandes Gratz and Norman Mintz, *Cities Back from the Edge: New Life for Downtown* (Hoboken, NJ: Wiley, 1998).
51. Eugénie Birch, "Downtown in the 'New American City,' " *Annals of the American Academy of Political and Social Science* 626 (2009): 149.
52. Rebecca Sohmer and Robert Lang, *Downtown Rebound* (Washington, DC: Fannie Mae Foundation / Brookings Institution, Center on Urban and Metropolitan Policy, 2001).
53. Alan Ehrenhalt, *The Great Inversion and the Future of the American City* (New York: Vintage Books, 2012).
54. Transportation Research Board, National Research Council, *Making Transit Work: Insight from Western Europe, Canada, and the United States—Special Report 257 (Washington, DC: National Academies Press, 2001).*
55. Jon Teaford. *The Twentieth-Century American City* (Baltimore: Johns Hopkins University Press, 1986).
56. American Public Transportation Association, "Record 10.8 Billion Trips Taken on U.S. Public Transportation in 2014," *Transit News,* March 9, 2015.

57. Richard Florida, 2010. "The Great Car Reset," *Atlantic Cities*, June 3, 2010.
58. Charles C. Tu and Mark J. Eppli, "An Empirical Examination of Traditional Neighborhood Developments," *Real Estate Economics* 29, no. 3 (2001): 485–501.
59. Center for Transit-Oriented Development, *Capturing the Value of Transit* (Berkeley, CA: Center for Transit-Oriented Development, 2008).
60. Belden Russonello & Stewart, *2004 National Community Preference Survey*, conducted for Smart Growth America and National Association of Realtors (Washington, DC: Belden Russonello & Stewart, 2004).
61. Belden Russonello & Stewart, *The 2011 Community Preference Survey: What Americans Are Looking for When Deciding Where to Live*, conducted for the National Association of Realtors (Washington, DC: Belden Russonello & Stewart, 2011).
62. Simmons Buntin, *Unsprawl: Remixing Spaces as Places* (Los Angeles: Planetizen Press, 2012).

CHAPTER 2

1. We have tried to use "United States" consistently rather than "America" to avoid the inference that somehow the United States and the American (or North American) continent are to be identified with each other. We have, however, used the adjectival version "American" where there appears to be no reasonable alternative to its use.
2. Kennedy wrote a book with that title. Romney used the phrase in his speech accepting the Republican presidential nomination in 2012. ABC News, "Mitt Romney's Speech at the Republican National Convention," August 30, 2012, http://abcnews.go.com/Politics/previewed-excerpts-mitt-romneys-speech-republican-national-convention/story?id=17120765.
3. Andrew H. Malcolm, *The Canadians* (New York: Times Books, 2005), xi.
4. Parliamentary Debates on the Subject of the Confederation of the British North American Provinces (Quebec: Hunter, Rose & Co., Parliamentary Printers, 1985), 32.
5. Quoted in J. Wheelwright, "Nationalism," United North America, August 8, 2005, http://www.unitednorthamerica.org/nationalism.htm.
6. Seymour Martin Lipset, *Continental Divide* (New York: Routledge, 1990), 53.
7. Kenneth McNaught, quoted in Seymour Martin Lipset, *Continental Divide* (New York: Routledge, 1990), 91. Lipset also quotes novelist Margaret Atwood that "Canada must be the only country in the world where a policeman is used as a national symbol," 90.
8. *The Canadian Encyclopedia*, s.v. "Canadian Identity," last edited December 15, 2013, http://www.thecanadianencyclopedia.ca/en/article/canadian-identity/.

9. G. R. Cook, "Canadian Centennial Celebrations," International Journal 12 (Autumn 1967): 663.
10. Rudyard Griffiths, interviewed in the *National Post*, December 28, 2012.
11. *The Canadian Encyclopedia*, s.v. "Canadian Identity."
12. Martin Patriquin, "The Epic Collapse of Quebec Separatism," *Maclean's*, April 11, 2014.
13. *Wikipedia*, s.v. "Newfoundland (Island)," last modified July 20, 2015, http://en.wikipedia.org/wiki/Newfoundland_(island).
14. What is referred to in Canada as the Aboriginal population includes three distinct subgroups: the Inuit, the First Nations, and the Métis. The last are descended from mixed Native and white ancestry; in contrast, however, to mixed populations in the United States, which are not perceived as a distinct ethnic group, the Métis both identify themselves and are recognized officially as a distinct part of the nation's Aboriginal population.
15. See, e.g., Lipset, *Continental Divide*, especially chap. 10, "Mosaic and Melting Pot."
16. Andrew Cohen, "Immigrants and Canadians, Maintaining Both Identities" *New York Times*, November 16, 2012. Cohen is the author of the highly critical 2003 best-seller *While Canada Slept: How We Lost Our Place in the World.*
17. Cohen, "Immigrants and Canadians."
18. A 2011 poll found that 92 percent of Canadians were satisfied or very satisfied with their lives, and a 2012 Gallup poll found that Canada is the second happiest country in the world, preceded only by Denmark. Interestingly, four out of the five Canadian metropolitan areas with the highest life satisfaction were in Quebec. Andrew Sharpe and Evan Capeluck, "Canadians Are Happy and Getting Happier: An Overview of Life Satisfaction in Canada, 2003–2011" (Ottawa: Center for the Study of Living Standards, 2012).
19. Equalization in Canada refers to the program by which federal revenues are distributed to the provinces on the basis of a formula that measures the disparity between each province's ability to raise revenues and equalizes the fiscal resources available to each province to provide services. Equalization payments are unconditional and are guaranteed by the Canadian Constitution of 1982.
20. Griffiths, interview.
21. Geert-Jan Hofstede and Michael Minkov, *Cultures and Organizations: Software of the Mind*, 3rd ed. (New York: McGraw-Hill, 2010), 6. The data that serve as the basis for the Hofstede model are based on integrating the findings from a number of different field surveys of residents of different countries; see *ibid*, 27–45. Unfortunately, copyright restrictions make it impossible for us to quote the actual country scores directly. Interested readers can find them at http://geert-hofstede.com/countries.html

22. Canadians are less likely to attend religious services than Americans, and when asked "how important to you are your religious beliefs" in a 1975 Gallup Poll, only 36 percent of Canadians answered "very important" compared with 56 percent of Americans. Lipset, *Continental Divide*, 84.
23. Michael Adams, *Fire and Ice: The United States, Canada and the Myth of Converging Values* (Toronto: Penguin Canada, 2004), 51.
24. Adams, *Fire and Ice*, 119. Adams cites a 2001 survey that found that 19 percent of Canadians owned guns.
25. Aaron Karp, *Small Arms Survey Research Notes No. 9: Estimating Civilian Owned Firearms* (Geneva: Graduate Institute of International and Development Studies, 2011).
26. Christian Boucher, "Canada-US Values Distinct, Inevitably Carbon Copy, or Narcissism of Small Differences?," *Horizons*, June 2014.
27. See generally Lipset, *Continental Divide*, 140–42, and Adams, *Fire and Ice*, 110–44; see also J. P. Alston, T. M. Morris, and A. Vedlitz, "Comparing Canadian and American Values: New Evidence from National Surveys" *American Review of Canadian Studies* 26, no. 3 (1996): 301–14.
28. There is considerable disagreement among sources about precisely what percentage of Canada's population lives within 100 miles of the US border, with estimates varying from 75 to 90 percent.
29. See http://data.worldbank.org/indicator/SP.URB.TOTL.IN.ZS, 2015.
30. The lower percentage for Canada is largely attributable to a narrower definition of metropolitan area used by Statistics Canada.
31. The World Bank, "GINI Index (World Bank Estimate)," 2015, http://data.worldbank.org/indicator/SI.POV.GINI.
32. For Canada, see Government of Canada, "Union Coverage in Canada, 2013," June 11, 2014, http://www.labour.gc.ca/eng/resources/info/publications/union_coverage/union_coverage.shtml; and for the United States, see Bureau of Labor Statistics, US Department of Labor, news release "Union Members—2014," January 23, 2015, http://www.bls.gov/news.release/pdf/union2.pdf. Both in Canada and the United States, however, public-sector workers are significantly more likely to be unionized than private-sector workers, although the disparity is less pronounced in Canada.
33. Data for Canada from Employment and Social Development Canada is from http://well-being.esdc.gc.ca/misme-iowb/.3ndic.1t.4r@-eng.jsp?iid=23; for the United States, from the Bureau of the Census,, https://www.census.gov/prod/2012pubs/acsbr11-01.pdf.
34. See, e.g., the research of the Broadbent Institute, at http://www.broadbentinstitute.ca/income_inequality.

35. Much of the data in this section come from the Central Intelligence Agency, "The World Factbook," https://www.cia.gov/library/publications/the-world-factbook/.
36. The measurement periods are slightly different. Canadian data are for the period 2001–2010, whereas US data are for the period 2000–2009.
37. Of course, 2 percent of the US population is a far larger number than 4 percent of the Canadian population.
38. Canada also has a Senate made up of members appointed by the governor-general on the prime minister's recommendation. Although in theory the two bodies are coequal, in practice the House of Commons is the dominant chamber, and it would not be too much of an exaggeration to characterize the Canadian system as having a de facto unicameral legislature.
39. The leader of the winning party is invited by the governor-general, who serves as ceremonial head of state representing the English monarch. Arguably, the power to select the prime minister and upon a prime minister's request to dissolve a government are the only substantive powers held by the governor-general.
40. Although it has never formed a national government, the NDP has at one time or another held the reins in five of Canada's eleven provinces. The NDP currently holds roughly 30 percent of the seats in the federal Parliament and has the status of official opposition.
41. Canada did not have a formal constitution until 1982.
42. In the same speech quoted earlier (see note 4), MacDonald decried the notion of "state's rights" as held in the United States and stressed that the proposed confederation would be radically different, noting that "we have strengthened the general government. We have given the general legislature all the great subjects of legislation" (33).
43. An interesting outcome of this particular separation of powers was the fate of the Ministry of State for Urban Affairs created in 1971 under the Trudeau government. Barred from exercising any direct authority in the realm of urban affairs, as Goldberg and Mercer write, "it quickly ran afoul of entrenched and power [national government] interests. It raised false expectations amongst the municipalities, and seemed to confirm provincial suspicions of larger federal role in urban development." It was abolished in 1979. Michael A. Goldberg and John Mercer, *The Myth of the North American City* (Vancouver: University of British Columbia Press, 1986), 123.
44. Quantifying this fact, however, is difficult. A handful of states are pure "Dillon's rule" states that provide little or no discretion to local governments, and a handful are pure home rule states; most, though, are a combination, including many like Illinois and Montana, which provide discretion to those municipalities that enact a charter or otherwise qualify as "home rule" municipalities.
45. Montana State Constitution, art. 11, § 6.

CHAPTER 3

1. For some of the metrics used below, we gathered our own data from national sources, making assumptions and adjusting the data as necessary to make the measures as meaningful as possible given different data collection and analysis practices in the two countries. For other measures, we relied on other researchers who had already done the difficult work of finding data and putting them into a comparable format.
2. We leave the economic development side of the equation to other researchers for the simple reason that it is outside our scope of expertise.
3. E. Fong, "A Comparative Perspective on Racial Residential Segregation: American and Canadian Experiences," *Sociological Quarterly* 37, no. 2 (1996): 199–226.
4. R. Alan Walks and Larry S. Bourne, "Ghettos in Canada's Cities? Racial Segregation, Ethnic Enclaves and Poverty Concentration in Canadian Urban Areas," *Canadian Geographer* 50, no. 3 (2006): 273–97.
5. It is interesting to note that segregation is less pronounced in Canadian than US cities even though Canadians tend to distinguish their own "mosaic" approach to immigration from the "melting pot" approach found south of the border. This ideological difference between the two countries is discussed in chapter 2.
6. Carlos Teixeira, Wei Li, and Audrey Kobayashi, eds., *Immigrant Geographies of North American Cities* (Don Mills, ON: Oxford University Press, 2011).
7. Zachary Taylor, "The Politics of Metropolitan Development: Institutions, Interests, and Ideas in the Making of Urban Governance in the United States and Canada, 1800–2000" (PhD diss., Graduate Department of Political Science, University of Toronto, 2015).
8. Thomas Gunton and K. S. Calbick, *The Maple Leaf in the OECD, Canada's Environmental Performance* (Vancouver: David Suzuki Foundation, 2010).
9. A. A. Sorensen, R. P. Greene, and K. Russ, *Farming on the Edge* (DeKalb, IL: American Farmland Trust, 1997).
10. Economist Intelligence Unit, 2011. *US and Canada Green City Index: Assessing the Environmental Performance of 27 Major US and Canadian Cities* (Munich: Siemens, 2011).
11. The metrics used in the report are not entirely consistent because they are drawn from public sources with different measurement approaches and with slight variations between metropolitan areas and countries. The authors of the report have chosen indicators and sources to make the comparisons as meaningful as possible.
12. Gokhan Egilmez, Serkan Gumus, and Murat Kucukvar, "Environmental Sustainability Benchmarking of the U.S. and Canada Metropoles: An Expert Judgment-Based Multi-Criteria Decision Making Approach," *Cities* 42 (2015): 31–41.

13. Grosvenor, *Resilient Cities: A Grosvenor Research Report*, 2014, http://www.grosvenor.com/getattachment/194bb2f9-d778-4701-a0ed-5cb451044ab1/ResilientCitiesResearchReport.pdf.
14. One indicator that might have been added is urban congestion, which touches on both environmental and livability aspects of sustainability. Canadian urbanites lose, on average, more time to traffic congestion than their US counterparts. We did not include this indicator because there is no agreement as to whether it is a positive or negative feature of Canadian cities. Traditionally, congestion has been considered a sign of urban dysfunction that must be overcome through greater investment in transportation infrastructure, usually roads and highways. Recently, however, views on this important issue have begun to change, and many urban and transportation planners are accepting congestion as not only a means of discouraging vehicle use and boosting demand for alternative modes, but as a sign of a successful, vibrant city.

CHAPTER 4

1. Joel Garreau, *Edge City: Life on the New Frontier* (New York: Anchor Books, 1992).
2. Northwest Environment Watch and Smart Growth BC, *Sprawl and Smart Growth in Greater Vancouver: A Comparison of Vancouver, British Columbia, with Seattle, Washington* (Seattle, WA: Northwest Environment Watch, 2002).
3. The relationship between density thresholds and transportation modes can be found in works such as Peter W. G. Newman and Jeffrey R. Kenworthy, *Cities and Automobile Dependence* (Brookfield. VT: Gower Technical Press, 1989).
4. D. Hernandez, M. Lister, and C. Suarez, *Location Efficiency and Housing Type: Boiling It Down to BTUs* (Washington, DC: US Environmental Protection Agency, 2011).
5. Reid Ewing and Robert Cervero, "Travel and the Built Environment: A Synthesis," *Transportation Research Record* 1780 (2001): 87–114.
6. John Miron, "Urban Sprawl in Canada and America: Just How Dissimilar?," paper presented at the annual meeting of the Association of American Geographers, New Orleans, LA, March 2003.
7. Zachary T. Taylor, "The Politics of Metropolitan Development: Institutions, Interests, and Ideas in the Making of Urban Governance in the United States and Canada, 1800–2000" (PhD diss., Graduate Department of Political Science, University of Toronto, 2015).
8. US Census Bureau, 2013 Housing Profile: United States. *American Housing Survey Factsheets*, May 2015; Statistics Canada, 2011 Canadian Census.

9. K. M. Leyden, "Social Capital and the Built Environment: The Importance of Walkable Neighborhoods," *American Journal of Public Health* 93, no. 9 (2003): 1546–51.
10. L. Frank, J. Sallis, T. Conway, J. Chapman, B. Saelens, and W. Bachman, "Many Pathways from Land Use to Health: Walkability Associations with Active Transportation, Body Mass Index, and Air Quality," *Journal of the American Planning Association* 72, no. 1 (2006): 75–87.
11. Gil Tal and Susan Handy, "Measuring Non-motorized Accessibility and Connectivity in a Robust Pedestrian Network," Final Research Report So2-2, UC Davis Sustainable Transportation Center.
12. https://www.walkscore.com.
13. Paul Torrens and Marina Alberti, *Measuring Sprawl* (London: Centre for Advanced Spatial Analysis, 2000).
14. Pierre Filion, Trudi Bunting, Kathleen McSpurren, and Alan Tse, "Canada-U.S. Metropolitan Density Patterns: Zonal Convergence and Divergence," *Urban Geography* 25, no. 1 (2004): 42–65.
15. Filion et al., "Canada-U.S. Metropolitan Density Patterns."
16. Peter Mieszkowski and Edwin Mills, "The Causes of Metropolitan Suburbanization," *Journal of Economic Perspectives* 7 (1993): 135–47.
17. Michael Lewyn, "Sprawl in Canada and The United States" (draft LLM thesis, University of Toronto, 2010).
18. Kim England and John Mercer, "Canadian Cities in Continental Context: Global and Continental Perspectives on Canadian Urban Development," in *Canadian Cities in Transition*, 3rd ed., eds. Trudi Bunting and Pierre Filion (Don Mills, ON: Oxford University Press, 2006), 24–39.
19. Yong Eun Shin, V. R. Vuchic, and E. Christian Bruun, "Land Consumption Impacts of a Transportation System on a City: An Analysis," *Transportation Research Record* 2110 (2009): 69–77.
20. Todd Litman, *Evaluating Transportation Land Use Impacts* (Victoria, BC: Victoria Transport Policy Institute 2015).
21. Robert A. Beauregard, *When America Became Suburban* (Minneapolis: University of Minnesota Press, 2006).
22. Peter W. G. Newman and Jeffrey R. Kenworthy, *Sustainability and Cities: Overcoming Automobile Dependence* (Washington DC: Island Press, 1999).
23. Reid Ewing, "Characteristics, Causes and Effects of Sprawl: A Literature Review," *Environmental and Urban Studies* 21, no. 2 (1994): 1–15.
24. Barry Edmonston, Michael A. Goldberg, and John Mercer, 1985. "Urban Form in Canada and the United States: An Examination of Urban Density Gradients," *Urban Studies* 22, no. 3 (1985): 209–17.

25. Tamim Raad, "The Car in Canada: A Study of Factors Influencing Automobile Dependence in Canada's Seven Largest Cities 1961–1991" (master's thesis, University of British Colombia, 1998).
26. Peter Newman and Jeffrey Kenworthy, *The End of Automobile Dependence: How Cities Are Moving beyond Car-Based Planning* (Washington, DC: Island Press, 2015).
27. The road measure includes all roads in each of the sample cities but does not account for variations in the number of lanes.
28. Robert Cervero et al., *Transit-Oriented Development in the United States: Experience, Challenges, and Prospects,* prepared for the Transportation Research Board of the National Research Council (Washington, DC: National Academies Press, 2004).
29. Newman and Kenworthy, *The End of Automobile Dependence.*
30. John Pucher and Ralph Buehler, "Why Canadians Cycle More Than Americans: A Comparative Analysis of Bicycling Trends and Policies," *Transport Policy* 13 (2006): 265–79.
31. John Pucher and Ralph Buehler, "International Overview," in *City Cycling,* eds. John Pucher and Ralph Buehler (Cambridge, MA: MIT Press, 2012), 9–30.
32. M. Winters, G. Davidson, D. Kao, and K. Teschke, "Motivators and Deterrents of Bicycling: Comparing Influences on Decisions to Ride," *Transportation* 38 (2011): 153–68.
33. John Pucher and Ralph Buehler, eds., *City Cycling* (Cambridge, MA: MIT Press, 2012).
34. John Pucher and Ralph Buehler, 2007. "Cycling in Canada and the United States: Why Canadians Are So Far Ahead," *Plan* (Spring/Summer 2007): 13–17.
35. P. Jacobsen, "Safety in Numbers: More Walkers and Bicyclists, Safer Walking and Bicycling," *Injury Prevention* 9 (2003): 205–9.
36. Paul Schimek, for one, has dismissed this factor as an explanation for differences in the mix of housing stock in the countries, noting that detached housing trends did not track income changes over time. See Paul Schimek, "Automobile and Public Transit Use in the United States and Canada: Comparison of Postwar Trends," *Transportation Research Record* 1521 (1996): 3–11.

CHAPTER 5

1. Michael A. Goldberg and John Mercer, *The Myth of the North American City: Continentalism Challenged* (Vancouver: University of British Columbia Press, 1986).

2. Donald Rothblatt, "North American Metropolitan Planning: Canadian and US Perspectives," *American Planning Association Journal* 60, no. 4 (1994): 501–20.
3. Paul Peterson, *City Limits* (Chicago: University of Chicago Press, 1981).
4. R. A. Rosenfeld and L. A. Reese, "Local Government Amalgamation from the Top Down," in *City-County Consolidation and Its Alternatives: Reshaping the Local Government Landscape*, eds. J. B. Carr and R. C. Feiock (Armonk, NY: M. E. Sharpe, 2004), 219–46.
5. S. Leland and K. Thurmaier, "City-County Consolidation: Do Governments Actually Deliver on Their Promises?," paper presented at the Urban Affairs Association Conference, Chicago, March 6–8, 2009.
6. Ruth DeHoog, D. Lowery, and W. Lyon, "Metropolitan Fragmentation and Suburban Ghettos: Some Empirical Observations on Institutional Racism," *Journal of Urban Affairs* 13, no. 4 (1991): 479–93.
7. Peter Dreier, John Mollenkopf, and Todd Swanstrom, *Place Matters: Metropolitics for the Twenty-First Century*, 3rd ed. (Lawrence: University Press of Kansas, 2014).
8. Anthony Perl and John Pucher, "Transit in Trouble? The Policy Challenge Posed by Canada's Changing Urban Mobility," *Canadian Urban Transit Association Forum*, June 5, 1995, 22–24.
9. For example, San Mateo and Marin counties have resisted the extension of the Bay Area Rapid Transit system into these populous parts of the San Francisco Bay area for decades. Interlocal haggling over routes and funding has prevented extension of the system to other parts of the region. As a result, the region does not have an integrated transit system. In Georgia, voters in suburban Cobb County blocked the extension of the Metropolitan Atlanta Regional Transit Agency rail service. In contrast, the extension of the subway and commuter rail systems into the suburbs around Toronto was welcomed by suburbanites.
10. Lawrence Katz and Kenneth Rosen, "The Interjurisdictional Effects of Growth Controls on Housing Prices," *Journal of Law and Economics* 30, no. 1 (1987): 149–60.
11. Anthony Downs, *New Visions for Metropolitan America* (Washington, DC: Brookings Institution / Lincoln Institute of Land Policy, 1994).
12. William Fulton, Rolf Pendall, Mai Nguyen, and Alicia Harrison, *Who Sprawls Most? How Growth Patterns Differ Across the US* (Washington, DC: Brookings Institution, Center on Urban and Metropolitan Policy, 2001).
13. Edward L. Glaeser, Matthew Kahn, and Chenghuan Chu, *Job Sprawl: Employment Location in U.S. Metropolitan Areas* (Washington, DC: Brookings Institution, Center on Urban and Metropolitan Policy 2001).
14. John I. Carruthers and Gudmundur F. Ulfarsson, "Fragmentation and Sprawl: Evidence from Interregional Analysis," *Growth and Change* 33, no. 3 (2002): 312–40.

15. John I. Carruthers, "Growth at the Fringe: The Influence of Political Fragmentation in United States Metropolitan Areas," *Papers in Regional Science* 82 (2003): 475–99.
16. Clinton v. Cedar Rapids and the Missouri River Railroad (24 Iowa 455; 1868).
17. Frances Frisken, *The Public Metropolis: The Political Dynamics of Urban Expansion in the Toronto Region, 1924–2003* (Toronto: Canadian Scholars' Press, 2007), 32.
18. The amalgamation of Metro Toronto with its lower-tier municipalities in 1998 was contested hotly by grassroot groups. A legal challenge failed on the grounds that a province has the authority to amalgamate cities in Canada, regardless of local sentiment. An appeal of the ruling was turned down by the Supreme Court of Canada, indicating that the lower court finding was justified.
19. Kenneth T. Jackson, *Crabgrass Frontier: The Suburbanization of the United States* (New York: Oxford University Press, 1987), 149.
20. David Rusk has argued that these "elastic cities" are less likely to be socially and economically segregated than inelastic cities that maintained their original boundaries despite suburban growth beyond their borders. David Rusk, *Cities without Suburbs* (Washington, DC: Woodrow Wilson Center Press, 1993).
21. John Cipman, *A Law unto Itself: How the Ontario Municipal Board Has Developed and Applied Land Use Planning Policy* (Toronto: University of Toronto Press, 2002).
22. Michael J. Skelly, *The Role of Canadian Municipalities in Economic Development* (Toronto: ICURR Press, 1995).
23. Laura Reese and Amy Malmer, "The Effects of State Enabling Legislation on Local Economic Development Policies," *Urban Affairs Quarterly* 30, no. 1 (1994): 114–35.
24. John M. Stevens and Robert P. McGowan, "Patterns and Predictors of Economic Development Power in Local Government: A Policy Perspective on Issues in One State," *Policy Studies Review* 6 (February 1987): 554–68.
25. Judith A. Garber and David L. Imbroscio, " 'The Myth of the North American City' Reconsidered: Local Constitutional Regimes in Canada and the United States," *Urban Affairs Review* 31, no. 5 (1996): 595–624.
26. Zachary Taylor and Neil Bradford, "The New Localism: Canadian Urban Governance in the Twenty-First Century," in *Canadian Cities in Transition: Perspectives for an Urban* Age, eds. P. Filion, M. Moos, T. Vinodrai, and R. Walker (Don Mills, ON: Oxford University Press, 2015), 194–208.
27. Ray Tomalty, *Innovative Infrastructure Financing Mechanisms for Smart Growth* (Vancouver: Smart Growth BC, 2007).
28. V. Ostrom, C. Tiebout, and R. Warren, "The Organization of Government in Metropolitan Areas: A Theoretical Inquiry," *American Political Science Review* 55 (1961): 831–42.

29. Myron Orfield, *Metropolitics: A Regional Agenda for Community and Stability* (Washington, DC: Brookings Institution Press, 1998).
30. William Fulton, *The Reluctant Metropolis* (Baltimore: Johns Hopkins University Press, 2001), 162–67.
31. Kurt Paulsen, "Geography, Policy or Market? New Evidence on the Measurement and Causes of Sprawl (and Infill) in U.S. Metropolitan Regions," *Urban Studies* 51, no. 12 (2014): 2629–45.
32. Ray Tomalty and Murtaza Haider, *Housing Affordability and Smart Growth in Calgary* (Calgary, AB: City of Calgary, 2008).
33. Andrew Sancton, "The Governance of Metropolitan Areas in Canada," *Public Administration and Development* 25, no. 4 (2005): 317–27.
34. M. Dann, *The Costs and/or Savings of Amalgamation* (Halifax, NS: HRM Amalgamation Project, 2004).
35. Sancton, "The Governance of Metropolitan Areas in Canada."
36. Caroline Andrew, "Evaluating Municipal Reform in Ottawa-Gatineau: Building for the Future," in *Metropolitan Governing: Canadian Cases, Comparative* Lessons, eds. Eran Razin and Patrick J. Smith (Jerusalem: Hebrew University Magnes Press, 2006), 75–94.
37. Cynthia Jacques, *Evaluating Smart Growth Efforts in Ottawa: A Report Card for the Ottawa 20/20 Official Plan* (supervised research project, master of urban planning degree, School of Urban Planning, McGill University, 2012.
38. Paul Tennant and David Zirnhelt, "Metropolitan Government in Vancouver: The Politics of Gentle Imposition," *Canadian Public Administration* 16 (Spring 1973): 128.
39. A. Artibise, K. Cameron, and J. H. Seelig, "Metropolitan Organization in Greater Vancouver: 'Do It Yourself' Regional Government," in *Metropolitan Governance without Metropolitan Government?*, ed. D. Phares (Burlington, VT: Ashgate Publishing, 2004), 195–211.
40. The GGH incorporates a number CMAs, including Toronto, Oshawa, Hamilton, Kitchener-Cambridge-Waterloo, Barrie, Saint Catharines–Niagara, Guelph, and Peterborough.
41. Maureen Carter-Whitney. *Ontario's Greenbelt in an International Context*, Friends of the Greenbelt Foundation Occasional Paper Series, No. 11 (Toronto: Canadian Institute for Environmental Law and Policy, 2010).
42. Henry Aubin, "A Failure on Almost All Counts: The Municipal Mergers Mean More Spending and More Bureaucrats, but Diminished Services," *Montreal Gazette*, March 3, 2006, A21.
43. Frances Frisken and Donald Rothblatt, 1994. "Summary and Conclusions," in *Metropolitan* Governance, eds. D. Rothblatt and A. Sancton (Berkeley: Institute of Governmental Studies Press, University of California), 433–66.

44. John Stuart Hall, "Who Will Govern American Metropolitan Regions, and How?," in *Governing Metropolitan Regions in the 21st Century*, ed. Don Phares (Armonk, NY: M. E. Sharpe, 2009), 62–87.
45. D. Wright, *Understanding Intergovernmental Relations* (Pacific Grove, CA: Brooks/Cole, 1988).
46. Ray Tomalty and Don Alexander, *Smart Growth in Canada: Implementation of a Planning Concept* (Ottawa: Canadian Mortgage and Housing Corporation, 2005).
47. Bernard Ross and Myron A. Levine, *Urban Politics: Cities and Suburbs in a Global Age* (Armonk, NY: M. E. Sharpe, 2012), 261.
48. David K. Hamilton, *Measuring the Effectiveness of Regional Governing Systems: A Comparative Study of City Regions in North America* (New York: Springer, 2014), 176.

CHAPTER 6

1. T. Moore, P. Thornes, and B. Appleyard, *The Transportation/Land Use Connection*, Planning Advisory Service Report Number 546/547 (Chicago: American Planning Association, 2007).
2. Christopher Fullerton, 2005. "A Changing of the Guard: Regional Planning in Ottawa, 1945–1974," *Urban History Review* 34, no. 1 (2005): 100–112.
3. Todd Litman, *Introduction to Multi-Modal Transportation Planning* (Victoria, BC: Victoria Transport Policy Institute, 2014).
4. Richard Oram, 1980. "The Role of Subsidy Policies in Modernizing the Structure of the Bus Transit Industry," *Transportation* 9, no. 4 (1980): 333–53.
5. Robert Cervero, "Urban Transit in Canada: Integration and Innovation at Its Best," *Transportation Quarterly* 40, no. 3 (1986): 293–316.
6. TransLink, "Making the Connection to a World of Choice," n.d.
7. The transit agency purchased the land in the planned rights-of-way, constructed the elevated tracks, and then sold the land under them to private investors who were permitted to transfer the development rights to station areas.
8. Robert Cervero, *The Transit Metropolis: A Global Inquiry* (Washington, DC: Island Press, 1998).
9. Donald Shoup, "Instead of Free Parking," *Access* 15, no. 2 (1999): 6–9.
10. Cervero, *The Transit Metropolis*.
11. J. Pucher, "Public Transport Developments: Canada vs. the United States," *Transportation Quarterly* 48, no. 1 (1994): 65–78.
12. Paul Mees, *Transport for Suburbia: Beyond the Automobile Age* (London: Earthscan, 2010).
13. Mees, *Transport for Suburbia*.

14. Cervero, *The Transit Metropolis*.
15. Paul Schimek, "Understanding the Relatively Greater Use of Public Transit in Canada Compared to the USA" (PhD diss., Urban Studies and Planning Department, Massachusetts Institute of Technology, 1997).
16. Paul Schimek, 1996. "Automobile and Public Transit Use in the United States and Canada: Comparison of Postwar Trends," *Transportation Research Record* 1521 (1996): 10.
17. Route miles do not take into account the far larger and continuous increase in lane miles (through widenings), interchanges, and so on.
18. Al Neuharth, "Traveling Interstates Is Our Sixth Freedom," *USA Today*, June 22, 2006.
19. Transportation Research Board, National Research Council, *Making Transit Work: Insight from Western Europe, Canada, and the United States—Special Report 257 (Washington, DC: National Academies Press, 2001).*
20. M. Wachs, "U.S. Transit Subsidy Policy: In Need of Reform," *Science*, June 1989, 1545–49.; and J. E. D. Richmond, *New Rail Transit Investments: A Review* (Cambridge, MA: John F. Kennedy School of Government, Harvard University, 1998).
21. Transportation Research Board, *Making Transit Work*.
22. Bruce Katz, Robert Puentes, and Scott Bernstein, "Getting Transportation Right for Metropolitan America," in *Taking the High Road: A Metropolitan Agenda for Transportation Reform*, eds. Bruce Katz and Robert Puentes (Washington, DC: Brookings Institution Press, 2005), 21–25.
23. Allison Padova, *Federal Participation in Highway Construction and Policy in Canada* (Ottawa: Library of Parliament, 2006).
24. Since 2005, about 40 percent of the approximately $5 billion collected annually from federal excise taxes has been directed into the now permanent Gas Tax Fund for municipal infrastructure. The money can be spent on almost any type of infrastructure, including public transit, wastewater infrastructure, drinking water, solid waste management, community energy systems, local roads, or highways.
25. Robert Cervero, "Urban Transit in Canada: Integration and Innovation at Its Best," *Transportation Quarterly* 40, no. 3 (1986): 293–316.
26. John Pucher and Christian Lefevre, *The Urban Transport Crisis in Western Europe and North America* (London: MacMillan, 1996).
27. Mees, *Transport for Suburbia*.
28. Todd Litman, *Transit Price Elasticities and Cross-Elasticities* (Victoria, BC: Victoria Transport Policy Institute, 2014).
29. James Howard Kunstler, *The Long Emergency: Surviving the Converging Catastrophes of the Twenty-First Century* (New York: Grove Press, 2005).

30. Pietro S. Nicola, *Laws of the Landscape: How Policies Shape Cities in Europe and America* (Washington, DC: Brookings Institution Press, 1999).
31. Transportation Research Board, *Special Report 251: Toward a Sustainable Future: Addressing the Long-Term Effects of Motor Vehicle Transportation on Climate and Ecology* (Washington, DC: National Research Council, 1997).
32. Peter Gordon and Bumsoo Lee, *Settlement Patterns in the U.S. and Canada: Similarities and Differences—Policies or Preferences?*, keynote address presented at the 26th Australasian Transport Research Forum Wellington, New Zealand, 2003.

CHAPTER 7

1. Rudyard Griffiths, interviewed in the *National Post*, December 28, 2012.
2. All dollar figures in this chapter are Canadian dollars except where clearly indicated to the contrary.
3. Equalization, as a specific constitutional mandate, is limited to the provinces. The federal government, however, also provides substantial financial transfers to the territories, at levels substantially higher in proportion to their population than the equalization payments going to the provinces.
4. Department of Finance Canada, "Equalization Program," last modified December 19, 2011, http://www.fin.gc.ca/fedprov/eqp-eng.asp.
5. Griffiths, interview.
6. Trudi Bunting, "Social Differentiation in Canadian Cities," in *Canadian Cities in Transition*, eds. Trudi Bunting and Pierre Filion (Toronto: Oxford University Press, 1991), 286–312.
7. Ontario Ministry of Finance, "Ontario Municipal Partnership Fund (OMPF)," last modified November 13, 2014, http://www.fin.gov.on.ca/en/budget/ompf/2015/.
8. Government of Saskatchewan, news release, "2014–15 Budget Continues Strong Support to Municipalities," March 19, 2014, http://www.finance.gov.sk.ca/budget2014-15/000-GRMunicipal.pdf.
9. Frances Frisken, "Canadian Cities and the American Example: A Prologue to Urban Policy Analysis," *Canadian Public Administration* 29, no. 3 (1986): 376.
10. The most well-known such case is in New Jersey, where a series of state supreme court decisions have led to a situation where the state now pays for 80 percent or more of the public education costs in a cluster of thirty-one designated "special needs" districts, primarily the state's impoverished (with respect to both fiscal capacity and the incomes of their residents) central cities. A more recent controversy is ongoing in the state of Kansas.

11. Rick Blizzard, "Healthcare System Ratings: U.S., Great Britain, Canada," Gallup Poll, March 25, 2003, http://www.gallup.com/poll/8056/healthcare-system-ratings-us-great-britain-canada.aspx.
12. "U.S. Health Care System—'Envy of the World'? Not in Canada!," Harris/Decima Poll, reported August 12, 2009, in http://www.news-medical.net/news/20090812/US-health-care-system-e2809cenvy-of-the-worlde2809d-not-in-Canada!.aspx.
13. Data from the World Health Organization, Global Health Observatory, http://www.who.int/gho/countries/en/. The years for which the data are provided vary between 2007 and 2012.
14. The LICO-AT is defined as the level at which a family spends 63.6 percent or more of its income after taxes on food, shelter, and clothing; see Citizens for Public Justice, *Poverty Trends Highlights 2013*.
15. In contrast to the United States, which has a single national "poverty" measure indifferent to metropolitan differences, StatisticsCanada calculates different LICO-AT figures based on community size. In 2012, the LICO-AT for a family of four ranged from $23,879 in rural nonmetropolitan areas to $36,504 in metropolitan areas with populations of more than 500,000.
16. The United States poverty rate for that year was defined as an income of US$22,350 for a family of four.
17. Canada without Poverty, "Poverty Progress Profiles," 2013 data, http://www.cwp-csp.ca/poverty/poverty-progress-profiles/.
18. Standing Committee on Human Resources, Skills and Social Development and the Status of Persons with Disabilities, House of Commons, *Federal Poverty Reduction Plan: Working in Partnership Towards Reducing Poverty in Canada* (Ottawa: Parliament of Canada, 2010).
19. It should be noted that these figures are for legal immigration and thus may exaggerate the difference between Canada and the United States with respect to *actual* immigration because it is likely that the United States has most accommodated a much larger number of undocumented immigrants than Canada, for fairly straightforward geographic reasons.
20. Nico Calavita and Alan Mallach, *Inclusionary Housing in International Perspective: Affordable Housing, Social Inclusion and Land Value Recapture* (Cambridge, MA: Lincoln Institute of Land Policy, 2010).
21. The housing finance system in the United States could have been characterized as a predominately private-sector one until its collapse in 2006 and 2007. Since then, as a result of the nationalization of Fannie Mae and Freddie Mac and the increased reliance on Federal Housing Administration mortgage insurance, it has

become fundamentally a public-sector driven system by default. Although virtually everyone involved in United States housing policy is in agreement that this situation should be changed, major disagreements between political parties, economic sectors, and interest groups over what form an alternative system should take have blocked change from taking place.

22. Canadian figure for 2011 from National Housing Survey; United States figure for 2013 from American Community Survey.
23. Canada Mortgage and Housing Corporation, "About CMHC," accessed January 11, 2015, https://www.cmhc-schl.gc.ca/en/corp/about/.
24. http://www.cmhc-schl.gc.ca/en/inpr/afhoce/fuafho/iah/index.cfm, accessed January 11, 2015.
25. See, generally, Edward L. Glaeser and Jesse M. Shapiro, "The Benefits of the Home Mortgage Interest Deduction," in *Tax Policy and the Economy*, NBER Book Series, vol. 17, ed. James Poterba (Cambridge, MA: MIT Press, 2003), 37–82.
26. Richard Voith, "Does the Federal Tax Treatment of Housing Affect the Pattern of Metropolitan Development?" *Federal Reserve Bank of Philadelphia Business Review* (1999): 3–16. Available at http://www.phil.frb.org/research-and-data/publications/business-review/1999/march-april/brma99rv.pdf.
27. Christopher Hume, "Big Ideas: Learning the lessons of St. Lawrence Neighborhood," *Toronto Star*, May 3, 2014.
28. Hume, "Big Ideas."
29. North Carolina, where countywide school districts are the norm, is a notable exception.
30. Gouvernement du Québec, Ministère de l'Éducation, du Loisir et du Sport, *Funding for Education in Québec at the Preschool, Elementary and Secondary School Levels: 2009–2010 School Year*, available at http://www.mels.gouv.qc.ca/dgfe/Financement/PDF/Funding2009_2010.pdf. The figures for Nova Scotia are very similar; in 2007, 77 percent of total school funding came from the province and 16 percent from municipal property taxation.
31. Juliana Herman, *Canada's Approach to School Funding: The Adoption of Provincial Control of Education Funding in Three Provinces* (Washington, DC: Center for American Progress, 2013).
32. Bruce Biddle and David Berliner, "A Research Synthesis: Unequal School Funding in the United States," *Beyond Instructional Leadership* 59, no. 8 (2002): 48–59.
33. Organization for Economic Cooperation and Development, "PISA 2012 Results," accessed January 12, 2015, http://www.oecd.org/pisa/keyfindings/pisa-2012-results.htm. The disparity in math scores noted in the text holds true in reading and science as well.

34. See in particular Peter Dreier, John Mollenkopf, and Todd Swanstrom (2001), *Place Matters: Metropolitics for the Twenty-First Century*, 3rd ed. (Lawrence: University Press of Kansas, 2001), especially chap. 2.

CHAPTER 8

1. 1951 data are from the Canada Year Book 1952–1953.
2. The only city shown in table 4-7 that can be considered to have even a remotely credible potential to regain its peak population within the next few decades is Washington, DC, which is currently estimated to have a population of not quite 660,000 and is adding nearly 10,000 people per year. It is an outlier among US cities, for many reasons.
3. The predictions of the imminent decline of many Sun Belt cities made by some planners and other analysts in the wake of the foreclosure crisis of 2006–2007 were not only premature, but tended to reflect more wishful thinking than solid analysis.
4. See Sean F. Reardon and Kendra Bischoff, "Growth in the Residential Segregation of Families by Income" (Providence, RI: US 2010 Project, Russell Sage Foundation and Brown University, 2011); and Kendra Bischoff and Sean F. Reardon, "Residential Segregation by Income, 1970–2009" (Providence, RI: US 2010 Project, Russell Sage Foundation and Brown University, 2013).
5. Because Canadian municipalities change boundaries frequently, the analysis compared 2001 and 2011 census tract data *only for those census tracts within the 2001 boundaries of the municipality*.
6. David Rusk, *Cities without Suburbs* (Washington, DC: Woodrow Wilson Center Press, 1993).
7. Chris Benner and Manuel Pastor, *Just Growth: Inclusion and Prosperity in America's Metropolitan Regions* (London: Routledge, 2012).
8. In 2010, the combined Chambers of Commerce and Boards of Trade in Canada issued a policy statement entitled "A Call for a National Urban Strategy Targeted Toward Canada's Largest Urban Centres" calling for fundamental reform of the structure of local government finance, writing that "the bottom line is that Canada's largest urban centers need a more stable, secure and growing revenue source. They need a new tax framework that does not increase the burden on taxpayers, but instead establishes a framework between the three levels of government that ensures tax revenues are sufficient to sustain municipal infrastructure and operations" (2). Available at http://halifaxchamber.com/wp-content/uploads/2015/05/A-Call-for-a-National-Urban-Strategy-Targeted-Toward-Canada%E2%80%99s-Largest-Urban-Centres-Nov-2010.pdf.

9. More US cities might well seek bankruptcy if they could; under US law, states have the power to control whether their constituent cities may file for bankruptcy and, if so, under what circumstances. Many states have passed laws effectively preventing cities from doing so; in a 2011 case, the state of Pennsylvania successfully intervened under such a law to have a bankruptcy petition filed by the city of Harrisburg dismissed.
10. *The Tyee*, an independent on-line newspaper in British Columbia, published an outstanding in-depth look at the current status of the False Creek South development and the challenges it faces in the future. See *The Tyee*, "False Creek South: An Experiment in Community," December 31, 2013, http://thetyee.ca/Series/2013/12/31/False-Creek-South-Experiment/.
11. Some of the retail vitality of downtown Chicago and San Francisco may well be a function of those two cities' vibrant tourism sector.
12. Specifically, as of 2013 the foreign-born percentages were Seattle 18 percent, Minneapolis 16 percent, and Washington, DC, 14 percent.
13. Although the literature is voluminous, two important sources are Wesley Skogan, *Disorder and Decline: Crime and the Spiral of Decay in American Neighborhoods* (Berkeley: University of California Press, 1990); and David S. Kirk and John H. Laub, "Neighborhood Change and Crime in the Modern Metropolis," *Crime and Justice* 39, no. 1 (2010): 441–502.
14. Nick Taylor-Vaisey, "Where Canadian Criminals Go to Play: A Look at the Cities with the Most Lawbreakers," *Maclean's*, November 29, 2012.
15. Ben Levin, "Comparing Canada and the U.S. on Education," *Education Week*, April 4, 2011, http://blogs.edweek.org/edweek/futures_of_reform/2011/04/comparing_canada_and_the_us_on_education.html.
16. New Haven Oral History Project, "Life in the Model City: Stories of Urban Renewal in New Haven," accessed March 3, 2105, http://www.yale.edu/nhohp/modelcity/process.html.
17. Kenneth Gosselin, "New Haven's 'Downtown Crossing' a Decade-Old Dream," *Hartford Courant*, July 6, 2012.
18. Mark Alden Branch, "Then ... and Now: How a City Came Back from the Brink," *Yale Alumni Magazine*, May/June 2009.
19. Gordon Stephenson, *A Redevelopment Study of Halifax, Nova Scotia* (Halifax, NS: City of Halifax, 1957).
20. Richard Bobier (1995) "Africville: The Test of Urban Renewal and Race in Halifax, Nova Scotia," *Past Imperfect* 4 (1995): 165.
21. *Wikipedia*, s.v. "Halifax (former city)," last modified July 9, 2015, http://en.wikipedia.org/wiki/Halifax_(former_city).
22. Jill Grant, "Hard Luck: The Failure of Regional Planning in Nova Scotia," *Canadian Journal of Regional Science* 12 (1989): 273–84.

23. Hugh Millward, "Peri-urban Residential Development in the Halifax Region 1960–2000: Magnets, Constraints and Planning Policies," *Canadian Geographer* 46, no. 1 (2002): 33–47.
24. Igor Vojnovic, "Municipal Consolidation in the 1990s: An Analysis of British Columbia, New Brunswick, and Nova Scotia," *Canadian Public Administration* 41, no. 2 (1998): 239–83.
25. Igor Vojnovic, "The Transitional Impacts of Municipal Consolidation," *Journal of Urban Affairs* 22, no. 4 (2000): 401.
26. Branch, "Then ... and Now."
27. New Haven Public Schools, "School Construction," accessed March 3, 2015, http://www.nhps.net/SchoolConstruction.
28. Jay F. May, "Connecticut," in *Charter School Funding: Inequity Expands*, eds. Jeagan Bardorff, Larry Maloney, Jay F. May, Sheree T. Speakman, Patrick J. Wolf, and Albert Cheng, April 2014, University of Arkansas, Department of Education Reform, http://www.uaedreform.org/wp-content/uploads/2014/charter-funding-inequity-expands-ct.pdf.
29. Halifax Regional Municipality, "The Greater Halifax Partnership—Economic Development Arm of HRM," May 2010, https://www.halifax.ca/IntergovernmentalAffairs/documents/TheGreaterHalifaxPartnership-EconomicDevelopmentArmofHRM.pdf.
30. Halifax Regional Municipality, "The Greater Halifax Partnership."
31. Jane Taber, "Interchange Demolition Plan Gives Halifax Renewal Rare Second Chance," *Globe and Mail*, May 8, 2014.
32. Statistics Canada, CANSIM, table 111-0009, last modified June 26, 2015, http://www5.statcan.gc.ca/cansim/a26?lang=eng&id=1110009. The Maritime provinces are New Brunswick, Nova Scotia, and Prince Edward Island.
33. All data come from the 2000 decennial census and the five-year 2009–2013 American Community Survey, available at http://factfinder.census.gov/faces/nav/jsf/pages/searchresults.xhtml?refresh=t#.
34. US Census Bureau, Center for Economic Studies, "On the Map," accessed March 3, 2015, http://onthemap.ces.census.gov/.
35. The New Haven–Milford Metropolitan Area, the city's region designated by the federal government, is considerably larger than the South Central Region designated by the state and contains a 2010 population of 862,000, more than twice that of the Halifax Regional Municipality.
36. South Central Regional Council of Governments, "About SCRCOG," accessed March 24, 2015, http://www.scrcog.org/who-we-are.html.

CHAPTER 9

1. Martin Turcotte, "The City/Suburb Contrast: How Can We Measure It?," *Canadian Social Trends* 85 (Summer 2008).
2. David Rusk, *Cities without Suburbs* (Washington DC: Woodrow Wilson Center Press, 1993, 9.
3. Michelle Wilde Anderson, "Dissolving Cities," *Yale Law Journal* 12 (2012): 1364–1446.
4. Given the absence of any central data source on such matters, it is impossible to prove whether that is actually the case. A Google search did find one municipality in Ohio that was ordered dissolved by court order. A notable recent case illustrating the problem was that of Hampton, Florida, a city with a population of 431, in which the mayor had been arrested for selling drugs and the municipal government was both largely incapable of providing services and notorious for illegal activity. Despite pressure on the state to dissolve the municipality, which would appear to be an obvious remedy for a dysfunctional situation, in the end, after finding (it is not clear on what basis) that the city had mended its ways, the state chose not to do so. One obstacle was that it was determined that dissolving the city would have required an act of the state legislature.
5. Robert C. Wood, *Suburbia: Its People and Their Politics* (Boston: Houghton Mifflin, 1958), 128.
6. Judith A. Garber and David L. Imbroscio, " 'The Myth of the North American City' Reconsidered: Local Constitutional Regimes in Canada and the United States," *Urban Affairs Review* 31, no. 5 (1996): 604.
7. Ontario Ministry of Infrastructure, *Growth Plan for the Greater Golden Horseshoe 2006, Office Consolidation 2013* (Toronto: Province of Ontario, 2013), available at https://www.placestogrow.ca/index.php?option=com_content&task=view&id=1&Itemid=8.
8. The Ontario Municipal Board has certain powers with respect to other municipal matters, such as the drawing of electoral district boundaries and the exercise of eminent domain or expropriation, but its agenda is dominated by land use and property matters. During the 2011–2012 fiscal year, the board received nearly two thousand appeals, of which nearly half came from the greater Toronto area.
9. The Ontario Municipal Board has been criticized for being too growth- and developer-oriented in its decisions.
10. Michael Lewyn, "Sprawl in Canada and the United States," *Urban Lawyer* 44, no. 1 (2012): 105.
11. Town of Lewiston, New York Code, accessed May 2, 2015, http://ecode360.com/27710559.

12. John Hasse, John Reiser, and Alexander Picharz, *Evidence of Exclusionary Effects of Land Use Policy within Historic and Projected Development Patterns in New Jersey: A Case Study of Monmouth and Somerset Counties* (Glassboro, NJ: Geospatial Research Laboratory, Rowan University, 2011).
13. Although this commitment has been significantly deemphasized by the current state administration, during the period covered by the study, however, it was being actively promoted by a series of state administrations.
14. Hasse, Reiser, and Picharz, *Evidence of Exclusionary Effects of Land Use Policy*.
15. Unfortunately, that was the lowest density category used in the study; observation of the areas studied makes clear that much of the development that took place was at even lower densities, such as 2-acre or 3-acre lots. The use of the term *rural* comes from the authors of the study, not the authors of this book.
16. Mia Baumeister, *Development Charges across Canada: An Underutilized Growth Management Tool?* (Toronto: Institute on Municipal Finance and Governance, University of Toronto, 2012).
17. Altus Group Economic Consulting, *Government Charges and Fees on New Homes in the Greater Toronto Area* (North York, ON: Building Industry and Land Development Association, 2013).
18. Clancy Mullen, *National Impact Fee Survey: 2012*, August 20, 2012, http://www.impactfees.com/publications%20pdf/2012_survey.pdf.
19. Pamela Blais, *Perverse Cities: Hidden Subsidies, Wonky Policy, and Urban Sprawl* (Vancouver: University of British Columbia Press, 2010).
20. Ray Tomalty, *Innovative Infrastructure Financing Mechanisms for Smart Growth* (Vancouver: Smart Growth BC, 2007).
21. *Northern Illinois Home Builders Ass'n v. County of DuPage*, 165 Ill.2d 25, 649 N.E.2d 384, 208 Ill.Dec. 328 (1995).
22. Jennifer Evans-Cowley, "Development Exactions: Process and Planning Issues," working paper (Cambridge, MA: Lincoln Institute of Land Policy, 2006), 8.
23. According to a study for the Brookings Institute, impact fees are imposed by 37 percent of jurisdictions in the top fifty metropolitan areas. See Rolf Pendall, Robert Puentes, and Jonathan Martin, *From Traditional to Reformed: A Review of the Land Use Regulations in the Nation's 50 Largest Metropolitan Areas* (Washington, DC: Brookings Institution Press, 2006).
24. Mullen, *National Impact Fee Survey*.
25. Arthur Nelson, Liza Bowles, Julian Juergensmeyer, and James Nicholas, *A Guide to Impact Fees and Housing Affordability* (Washington, DC: Island Press, 2008).
26. F. Kaid Benfield, Matthew Raimi, and Donald Chen, *Once There Were Greenfields: How Urban Sprawl is Undermining America's Environment, Economy, and Social Fabric* (Washington, DC: Natural Resources Defense Council, 1999).

27. Reid Ewing, "Coordinating Land Use and Transportation in Sacramento," *Planning*, April 2013, 52. The character of MPOs varies widely. Some states gave this responsibility to existing multifaceted regional planning agencies, whereas others created or designated single-purpose transportation planning entities. In some states, the MPOs have some degree of independence, whereas in others, they are little more than adjuncts of the state transportation department. Their lack of authority over local planning, however, is all but universal.
28. This estimate is ours; no single figure for total office space along the corridor (or uniform definition of the corridor) exists. Two major complexes, Carnegie Center and Forrestal Center, each contain more than 2 million square feet of leasable space.
29. Although the project still appears on the New Jersey DOT website, it is clear that it has not been updated recently; it refers to the final report "expected to be completed in 2011." See State of New Jersey Department of Transportation, "New Jersey FIT: Future in Transportation," last updated March 15, 2011, http://www.state.nj.us/transportation/works/njfit/route1rsg.shtm.
30. Although Markham is a "city" and Huntington is a "town," these differences in wording are terminology alone and do not reflect any differences in the scope of local authority.
31. American Institute of Certified Planners, the national body that certifies professional city planners in the United States.
32. Town of Huntington, New York, *Horizons 2020: A Comprehensive Plan Update*, December 2008, http://www.huntingtonny.gov/filestorage/13749/13847/16804/16874/16878/20879/872sm.pdf.
33. Town of Huntington, New York, iv–v.
34. Town of Huntington, New York, I-7.
35. Town of Huntington, New York, VI-8, VI-10.
36. Areas shown in dark brown on the map are incorporated villages, which, although considered to be within the town from a legal standpoint, adopt their own comprehensive plans and zoning ordinances.
37. City of Markham, Ontario. *Markham Official Plan, Planning Markham's Future*, 2013, 1-3.
38. *Markham Official Plan, Framework Element*, 2-15. Assuming an average household size of 2.5, that is roughly comparable to a minimum residential density of 32 dwelling units per acre.
39. *Markham Official Plan, Land Use Element*, 8-15.
40. *Markham Official Plan, Framework Element*, 2-12.
41. *Markham Official Plan, Framework Element*, 2-12.
42. City of Brossard, Quebec, *Planification Stratégique 2013–2030*, 2013, 11.
43. Brossard, *Planification Stratégique 2013–2030*, 22.

44. Brossard, *Planification Stratégique 2013–2030*, 29.
45. Brossard, *Planification Stratégique 2013–2030*, 5.

CHAPTER 10

1. "Buy a House, Get a Streetcar" Better! Cities and Towns, October/November 2012, http://bettercities.net/article/buy-house-get-streetcar-19120.
2. Jason Miller, "Another Greyfield Gone: Belmar in Lakewood, Colorado," *Town Paper*, Fall 2005.
3. American Policy Center, "Agenda 21," accessed April 18, 2015, http://american policy.org/agenda21/. Glenn Beck, a perennial right-wing media figure, has published a dystopian novel entitled *Agenda 21: Into the Shadows*, purporting to describe the world in which, as described on his website, "following the worldwide implementation of a UN-led program called Agenda 21, the once-proud people of America have become obedient residents who live in barren, brutal Compounds and serve the autocratic, merciless Authorities." Glenn Beck, accessed April 18, 2015, http://www.glennbeck.com /agenda21/?utm_source=glennbeck&utm_medium=contentcopy_link.
4. Linda McIntyre, "Agenda 21 and the Planning Commission," *Commissioner*, Summer 2013, 1–2.
5. Urban Land Institute, *Shifting Suburbs: Reinventing Infrastructure for Compact Development* (Washington, DC: Urban Land Institute, 2012), 2.
6. The jurisdiction of the Massachusetts Housing Appeals Committee is narrow; namely, it is to determine whether proposals for developments that include affordable housing have been unreasonably denied or approved subject to unreasonable conditions. In comparison with other states where such matters are left to the court system, however, it has demonstrated that such an administrative body can provide substantially more effective and timely relief.
7. Michael Anderson, "Selling Biking: Perceived Safety, the Barrier That Still Matters," *People for Bikes*, November 7, 2013.
8. For a good description of the background and controversy around Michigan's emergency financial management law, see Ryan Holywell, "Emergency Financial Managers: Michigan's Unwelcome Savior," *Governing*, May 2012, 34–40. It is worth noting that in many cases, the protest does not stem from objections to the exercise of state power in principle as much as the perceived or real concern that the state is intervening on behalf of bondholders or creditors rather than in the broader public interest.
9. Under a 1963 law, California created county-level local area formation commissions, which have the power to form, dissolve, and change the boundaries of special districts as well as to approve annexation of unincorporated areas by

municipalities. Although it can also recommend creation or dissolution, or changes to boundaries of general-purpose municipalities, those changes must be approved by the voters of the affected municipalities.

10. BRAC is not a standing commission as its name would imply, but a series of separate commissions, each one authorized separately by Congress. There have been five BRAC authorizations since the first one, in 1988.
11. An effort along these lines was made in New Jersey, which led to the establishment of the Local Unit Alignment, Reorganization and Consolidation Commission (LUARCC) by state law in 2007. Although the initial legislative proposal provided that LUARCC's recommendations would be effectuated by an up-or-down vote of the state legislature, without regard to local wishes, it was amended before passage to provide that recommendations would only go into effect, *after* such a vote, if approved by a majority of the voters in each affected municipality. Although still technically in existence, a review of the agency website indicates that it has no staff and has not met since 2011.
12. Lawrence O. Picus and Associates, *An Evaluation of Vermont's Education Finance System* (Montpelier: Vermont Joint Fiscal Office, 2012), available at http://www.leg.state.vt.us/jfo/Education%20RFP%20Page/Picus%20and%20Assoc%20VT%20Finance%20Study%20with%20Case%20Studies%201-2-12a.pdf.
13. The Vermont Supreme Court found that that state's constitution required that all the state's school pupils were entitled to equal educational opportunity, which was violated by the then-extant school funding formula; see *Brigham v. State of Vermont*, 166 VT 246. The New Jersey state constitution, for example, provides (sec. IV, pt. 1) that "the Legislature shall provide for the maintenance and support of a thorough and efficient system of free public schools for the instruction of all the children in the State between the ages of five and eighteen years," which was the basis for that state's supreme court requiring fundamental changes to the state's program of aid to urban school districts.
14. One example, found in New York and Ohio and possibly elsewhere, are state laws providing that the state is responsible for maintaining state roads outside cities, but that inside cities, municipal governments have that responsibility. These laws are holdovers from the nineteenth century when cities were wealthy and their surrounding rural areas poor.
15. 512 US 687, 12. It should be noted that in *Dolan*, the issue at hand was dedication of part of the property for public open space rather than a cash contribution, but the principle is the same. It should be noted that the court did not object to the proposition of requiring dedication of open space, but found the city's requirement to be vastly out of proportion to any possible effect of Dolan's proposal, which was for the expansion of a modest electrical and plumbing supply store.

16. Under federal law initially enacted in 1986, each state receives a formula amount of federal tax credits that it can allocate to developers of low-income rental housing. Because the demand for credits generally exceeds the amount that states have available to allocate, each state is required to adopt annually a Qualified Allocation Plan that sets forth a point scoring system by which it determines which proposals receive allocations. States have very broad discretion to determine what criteria are used to allocate their credits; thus, a preference of this sort would be very easy to incorporate into their procedures.
17. John Norquist, "Roadblock on Main Street: How Federal Finance Policies Run Downtowns Out of Business," *American Conservative*, November/December 2014.

Bibliography

Adams, Michael. 2004. *Fire and Ice: The United States, Canada and the Myth of Converging Values.* Toronto: Penguin Canada.

Alston, J. P., T. M. Morris, and A. Vedlitz. 1996. "Comparing Canadian and American Values: New Evidence from National Surveys." *American Review of Canadian Studies* 26, no. 3: 301–14.

Altus Group Economic Consulting. 2013. *Government Charges and Fees on New Homes in the Greater Toronto Area.* North York, ON: Building Industry and Land Development Association.

American Public Transportation Association. 2015. "Record 10.8 Billion Trips Taken on U.S. Public Transportation in 2014." *Transit News,* March 9.

American Society of Civil Engineers. 2013. *2013 Report Card for America's Infrastructure.* Reston, VA: American Society of Civil Engineers.

Anderson, Michael. 2013. "Selling Biking: Perceived Safety, the Barrier That Still Matters." *People for Bikes,* November 7.

Anderson, Michelle Wilde. 2012. "Dissolving Cities." *Yale Law Journal* 12: 1364–1446.

Andrew, Caroline. 2006. "Evaluating Municipal Reform in Ottawa-Gatineau: Building for the Future." In *Metropolitan Governing: Canadian Cases, Comparative Lessons,* edited by Eran Razin and Patrick J. Smith, 75–94. Jerusalem: Hebrew University Magnes Press.

Arroyo, Carmen. 2008. *The Funding Gap.* Washington, DC: Education Trust.

Artibise, K. Cameron, and J. H. Seelig. 2004. "Metropolitan Organization in Greater Vancouver: 'Do It Yourself' Regional Government." In *Metropolitan Governance without Metropolitan Government?,* edited by D. Phares, 195–211. Burlington, VT: Ashgate.

Aubin, Henry. 2006. "A Failure on Almost All Counts: The Municipal Mergers Mean More Spending and More Bureaucrats, but Diminished Services." *Montreal Gazette*, March 3, A21.

Baumeister, Mia. 2012. *Development Charges across Canada: An Underutilized Growth Management Tool?* Toronto: Institute on Municipal Finance and Governance, University of Toronto.

Beauregard, Robert A. 2006. *When America Became Suburban*. Minneapolis: University of Minnesota Press.

Belden Russonello & Stewart. 2004. *2004 National Community Preference Survey*. Conducted for Smart Growth America and National Association of Realtors. Washington, DC: Belden Russonello & Stewart.

———. 2011. *The 2011 Community Preference Survey: What Americans Are Looking for When Deciding Where to Live*. Conducted for the National Association of Realtors. Washington, DC: Belden Russonello & Stewart.

Benfield, F. Kaid, Matthew Raimi, and Donald Chen. 1999. *Once There Were Greenfields: How Urban Sprawl Is Undermining America's Environment, Economy, and Social Fabric*. Washington, DC: Natural Resources Defense Council.

Benner, Chris, and Manuel Pastor. 2012. *Just Growth: Inclusion and Prosperity in America's Metropolitan Regions*. London: Routledge.

———. 2015. "Brother, Can You Spare Some Time? Sustaining Prosperity and Social Inclusion in America's Metropolitan Regions." *Urban Studies* 52, no. 7: 1339–56.

Biddle, Bruce, and David Berliner. 2002. "A Research Synthesis: Unequal School Funding in the United States." *Beyond Instructional Leadership* 59, no. 8: 48–59.

Birch, Eugénie. 2009. "Downtown in the 'New American City.'" *Annals of the American Academy of Political and Social Science* 626: 134–53.

Bischoff, Kendra, and Sean F. Reardon. 2013. *Residential Segregation by Income, 1970–2009*. Providence, RI: US 2010 Project, Russell Sage Foundation and Brown University.

Blais, Pamela. 2010. *Perverse Cities: Hidden Subsidies, Wonky Policy, and Urban Sprawl*. Vancouver: University of British Columbia Press.

Bobier, Richard. 1995. "Africville: The Test of Urban Renewal and Race in Halifax, Nova Scotia." *Past Imperfect* 4: 163–80.

Boucher, Christian. 2014. "Canada-US Values Distinct, Inevitably Carbon Copy, or Narcissism of Small Differences?" *Horizons*, June.

Branch, Mark Alden. 2009. "Then ... and Now: How a City Came Back from the Brink." *Yale Alumni Magazine*, May/June.

Bruegmann, Robert. 2005. *Sprawl: A Compact History*. Chicago: University of Chicago Press.

Buntin, Simmons. 2012. *Unsprawl: Remixing Spaces as Places*. Los Angeles: Planetizen Press.

Bunting, Trudi. 1991. "Social Differentiation in Canadian Cities." In *Canadian Cities in Transition*, edited by Trudi Bunting and Pierre Filion, 286–312. Toronto: Oxford University Press.

Burchell, Robert, et. al. 1998. *The Costs of Sprawl—Revisited.* Prepared for the Transportation Research Board of the National Research Council. Washington, DC: National Academies Press.

Burchell, Robert, G. Lowenstein, W. R. Dolphin, C. C. Galley, A. Downs, S. Seskin, K. G. Still, P. Brinckerhoff, and T. Moore. 2002. *Costs of Sprawl—2000.* Washington, DC: Transportation Research Board.

Calavita, Nico, and Alan Mallach. 2010. *Inclusionary Housing in International Perspective: Affordable Housing, Social Inclusion and Land Value Recapture*. Cambridge, MA: Lincoln Institute of Land Policy.

Cambridge Systematics, Inc. 2009. *Moving Cooler: An Analysis of Transportation Strategies for Reducing Greenhouse Gas Emissions.* Washington, DC: Urban Land Institute.

Carruthers, John I. 2003. "Growth at the Fringe: The Influence of Political Fragmentation in United States Metropolitan Areas." *Papers in Regional Science* 82: 475–99.

Carruthers, John I., and Gudmundur F. Úlfarsson. 2002. "Fragmentation and Sprawl: Evidence from Interregional Analysis." *Growth and Change* 33, no. 3: 312–40.

Carter-Whitney, Maureen. 2010. *Ontario's Greenbelt in an International Context.* Friends of the Greenbelt Foundation Occasional Paper Series, No. 11. Toronto: Canadian Institute for Environmental Law and Policy.

Center for Transit-Oriented Development. 2008. *Capturing the Value of Transit.* Berkeley, CA: Center for Transit-Oriented Development.

Cervero, Robert. 1986. "Urban Transit in Canada: Integration and Innovation at Its Best." *Transportation Quarterly* 40, no. 3: 293–316.

———. 1998. *The Transit Metropolis: A Global Inquiry.* Washington, DC: Island Press.

Cervero, Robert, et al. 2004. *Transit-Oriented Development in the United States: Experience, Challenges, and Prospects.* Prepared for the Transportation Research Board of the National Research Council. Washington, DC: National Academies Press.

Champion, L. 1990. "The Relationship between Social Vulnerability and the Occurrence of Severely Threatening Life Events." *Psychological Medicine* 20, no. 1: 157–61.

Cipman, John. 2002. *A Law unto Itself: How the Ontario Municipal Board Has Developed and Applied Land Use Planning Policy*. Toronto: University of Toronto Press.

City of Brossard, Quebec. 2013. *Planification Stratégique 2013–2030*.

Cohen, Andrew. 2012. "Immigrants and Canadians, Maintaining Both Identities." *New York Times*, November 16.

Cook, G. R. 1967. "Canadian Centennial Celebrations." *International Journal* 12 (Autumn): 659–63.

Dann, M. 2004. *The Costs and/or Savings of Amalgamation*. Halifax, NS: Halifax Amalgamation Project.

Dannenberg, Andrew, Howard Frumkin, and Richard Jackson, eds. 2011. *Making Healthy Places: Designing and Building for Health, Well-Being, and Sustainability*. Washington, DC: Island Press.

DeHoog, Ruth, D. Lowery, and W. Lyon. 1991. "Metropolitan Fragmentation and Suburban Ghettos: Some Empirical Observations on Institutional Racism." *Journal of Urban Affairs* 13, no. 4: 479–93.

Downs, Anthony. 1994. *New Visions for Metropolitan America*. Washington, DC: Brookings Institution / Lincoln Institute of Land Policy.

Dreier, Peter, John Mollenkopf, and Todd Swanstrom. 2014. *Place Matters: Metropolitics for the Twenty-First Century*. 3rd ed. Lawrence: University Press of Kansas.

Economist Intelligence Unit. 2011. *US and Canada Green City Index: Assessing the Environmental Performance of 27 Major US and Canadian Cities*. Munich: Siemens.

Edmonston, Barry, Michael A. Goldberg, and John Mercer. 1985. "Urban Form in Canada and the United States: An Examination of Urban Density Gradients." *Urban Studies* 22, no. 3: 209–17.

Egilmez, Gokhan, Serkan Gumus, and Murat Kucukvar. 2015. "Environmental Sustainability Benchmarking of the U.S. and Canada Metropoles: An Expert Judgment-Based Multi-Criteria Decision Making Approach." *Cities* 42: 31–41.

Ehrenhalt, Alan. 2012. *The Great Inversion and the Future of the American City*. New York: Vintage Books.

England, Kim, and John Mercer. 2006. "Canadian Cities in Continental Context: Global and Continental Perspectives on Canadian Urban Development." In *Canadian Cities in Transition*, 3rd ed., edited by Trudi Bunting and Pierre Filion, 24–39. Don Mills, ON: Oxford University Press.

Evans-Cowley, Jennifer. 2006. "Development Exactions: Process and Planning Issues." Working paper. Lincoln Institute of Land Policy, Cambridge, MA.

Ewing, R., R. Schieber, and C. Zegeer. 2003. "Urban Sprawl as a Risk Factor in Motor Vehicle Occupant and Pedestrian Fatalities." *American Journal of Public Health* 93, no.9: 1541–45.

Ewing, Reid. 1994. "Characteristics, Causes and Effects of Sprawl: A Literature Review." *Environmental and Urban Studies* 21, no. 2: 1–15.

———. 2013. "Coordinating Land Use and Transportation in Sacramento." *Planning*, April, 52–53.

Ewing, Reid, and Robert Cervero. 2001. "Travel and the Built Environment—Synthesis." *Transportation Research Record* 1780: 87–112.

Ewing, Reid, and Shima Hamidi. 2014. *Measuring Urban Sprawl and Validating Sprawl Measures.* Washington, DC: National Institutes of Health and Smart Growth America.

Filion, Pierre, Trudi Bunting, Kathleen McSpurren, and Alan Tse. 2004. "Canada-U.S. Metropolitan Density Patterns: Zonal Convergence and Divergence." *Urban Geography* 25, no. 1: 42–65.

Florida, Richard. 2004. *The Rise of the Creative Class.* St. Louis, MO: Turtleback Books.

———. 2010. "The Great Car Reset." *Atlantic Cities*, June 3.

Fong, E. 1996. "A Comparative Perspective on Racial Residential Segregation: American and Canadian Experiences." *Sociological Quarterly* 37, no. 2: 199–226.

Frank, L., J. Sallis, T. Conway, J. Chapman, B. Saelens, and W. Bachman. 2006. "Many Pathways from Land Use to Health: Walkability Associations with Active Transportation, Body Mass Index, and Air Quality." *Journal of the American Planning Association* 72, no. 1: 75–87.

Frank, Lawrence, and Peter Engelke. 2005. "Multiple Impacts of the Built Environment on Public Health: Walkable Places and the Exposure to Air Pollution." *International Regional Science Review* 28: 193–216.

Frank, Lawrence, Brian Stone, and William Bachman. 2000. "Linking Land Use with Household Vehicle Emissions in the Central Puget Sound: Methodological Framework and Findings." *Transportation Research Part D* 5, no. 3: 173–96.

Frisken, Frances. 1986. "Canadian Cities and the American Example: A Prologue to Urban Policy Analysis." *Canadian Public Administration* 29, no. 3: 345–76.

———. 2007. *The Public Metropolis: The Political Dynamics of Urban Expansion in the Toronto Region, 1924–2003.* Toronto: Canadian Scholars' Press.

Frisken, Frances, and Donald Rothblatt. 1994. "Summary and Conclusions." In *Metropolitan Governance*, edited by D. Rothblatt and A. Sancton, 433–66. Berkeley: Institute of Governmental Studies Press, University of California.

Fullerton, Christopher. 2005. "A Changing of the Guard: Regional Planning in Ottawa, 1945–1974." *Urban History Review* 34, no. 1: 100–112.

Fulton, William, Rolf Pendall, Mai Nguyen, and Alicia Harrison. 2001. *Who Sprawls Most? How Growth Patterns Differ across the US.* Washington, DC: Brookings Institution, Center on Urban and Metropolitan Policy.

Fulton, William. 2001. *The Reluctant Metropolis.* Baltimore: Johns Hopkins University Press.

Garber, Judith A., and David L. Imbroscio. 1996. " 'The Myth of the North American City' Reconsidered: Local Constitutional Regimes in Canada and the United States." *Urban Affairs Review* 31, no. 5: 595–624.

Garreau, Joel. 1992. *Edge City: Life on the New Frontier.* New York: Anchor Books.

Glaeser, Edward L., Matthew Kahn, and Chenghuan Chu. 2001. *Job Sprawl: Employment Location in U.S. Metropolitan Areas.* Washington, DC: Brookings Institution, Center on Urban and Metropolitan Policy.

Glaeser, Edward L., and Jesse M. Shapiro. 2003. "The Benefits of the Home Mortgage Interest Deduction." In *Tax Policy and the Economy*, NBER Book Series, vol. 17, edited by James Poterba, 37–82. Cambridge, MA: MIT Press.

Goldberg, Michael A., and John Mercer. 1986. *The Myth of the North American City: Continentalism Challenged.* Vancouver: University of British Columbia Press.

Goonetilleke, A., E. Thomas, S. Ginn, and D. Gilbert. 2005. "Understanding the Role of Land Use in Urban Stormwater Quality Management." *Journal of Environmental Management* 74: 31–42.

Gordon, Peter, and Bumsoo Lee. 2003. *Settlement Patterns in the US and Canada: Similarities and Differences—Policies or Preferences?* Keynote address presented at the 26th Australasian Transport Research Forum, Wellington, New Zealand.

Gosselin, Kenneth 2012. "New Haven's 'Downtown Crossing' a Decade-Old Dream." *Hartford Courant*, July 6.

Gouvernement du Québec, Ministére de l'Éducation, du Loisir et du Sport. 2009. *Funding for Education in Québec at the Preschool, Elementary and Secondary School Levels: 2009–2010 School Year.*

Grant, Jill. 1989. "Hard Luck: The Failure of Regional Planning in Nova Scotia." *Canadian Journal of Regional Science* 12: 273–84.

Gratz, Roberta Brandes, and Norman Mintz. 1998. *Cities Back from the Edge: New Life for Downtown.* Hoboken, NJ: Wiley.

Grosvenor. 2014. *Resilient Cities: A Grosvenor Research Report.*

Gunton, Thomas, and K. S. Calbick. 2010. *The Maple Leaf in the OECD, Canada's Environmental Performance.* Vancouver: David Suzuki Foundation.

Hall, John Stuart. 2009. "Who Will Govern American Metropolitan Regions, and How?" In *Governing Metropolitan Regions in the 21st Century*, edited by Don Phares, 62–87. Armonk, NY: M. E. Sharpe.

Hamilton, David K. 2014. *Measuring the Effectiveness of Regional Governing Systems: A Comparative Study of City Regions in North America.* New York: Springer.

Hasse, John, John Reiser, and Alexander Picharz. 2011. *Evidence of Exclusionary Effects of Land Use Policy within Historic and Projected Development Patterns in New Jersey: A Case Study of Monmouth and Somerset Counties.* Glassboro, NJ: Geospatial Research Laboratory, Rowan University.

Herman, Juliana. 2013. *Canada's Approach to School Funding: The Adoption of Provincial Control of Education Funding in Three Provinces.* Washington, DC: Center for American Progress.

Hernandez, D., M. Lister, and C. Suarez. 2011. *Location Efficiency and Housing Type: Boiling It Down to BTUs.* Washington, DC: US Environmental Protection Agency.

Hofstede, Geert, Geert-Jan Hofstede, and Michael Minkov. 2010. *Cultures and Organizations: Software of the Mind.* 3rd ed. New York: McGraw-Hill.

Holywell, Ryan 2012. "Emergency Financial Managers: Michigan's Unwelcome Savior." *Governing,* May: 34–40.

Hume, Christopher. 2014. "Big Ideas: Learning the Lessons of St. Lawrence Neighborhood." *Toronto Star,* May 3.

Jackson, Kenneth T. 1987. *Crabgrass Frontier: The Suburbanization of the United States.* New York: Oxford University Press.

Jacobsen, P. 2003. "Safety in Numbers: More Walkers and Bicyclists, Safer Walking and Bicycling." *Injury Prevention* 9: 205–9.

Karp, Aaron. 2011. *Small Arms Survey Research Notes No. 9: Estimating Civilian Owned Firearms*. Geneva: Graduate Institute of International and Development Studies.

Katz, Bruce, Robert Puentes, and Scott Bernstein. 2005. "Getting Transportation Right for Metropolitan America." In *Taking the High Road: A Metropolitan Agenda for Transportation Reform,* edited by Bruce Katz and Robert Puentes, 21–25. Washington, DC: Brookings Institution Press.

Katz, Bruce, and Joel Rogers. 2001. "The Next Urban Agenda." In *The Next Agenda: Blueprint for a New Progressive Movement,* edited by R. Borosage and R. Hickey, 189–210. Boulder, CO: Westview.

Katz, Lawrence, and Kenneth Rosen. 1987. "The Interjurisdictional Effects of Growth Controls on Housing Prices." *Journal of Law and Economics* 30, no. 1: 149–60.

Kennedy, C., J. Steinberger, B. Gasson, Y. Hansen, T. Hillman, M. Havranek, D. Pataki, A. Phdungsilp, A. Ramaswami, and G. Villalba Mendez. 2009. "Greenhouse Gas Emissions from Global Cities." *Environmental Science and Technology* 43: 7297–7302.

Kirk, David S., and John H. Laub. 2010. "Neighborhood Change and Crime in the Modern Metropolis." *Crime and Justice* 39, no. 1: 441–502.

Kneebone, Elizabeth, and Alan Berube. 2013. *Confronting Suburban Poverty in America*. Washington, DC: Brookings Institution Press.

Kunstler, James Howard. 2005. *The Long Emergency: Surviving the Converging Catastrophes of the Twenty-First Century*. New York: Grove Press.

Lawrence O. Picus and Associates. 2012. *An Evaluation of Vermont's Education Finance System*. Montpelier: Vermont Joint Fiscal Office.

Lawton, Keith. 2001. "The Urban Structure and Personal Travel: An Analysis of Portland, Oregon Data and Some National and International Data." Paper presented at the E-Vision 2000 Conference, Washington, DC, October 11–13.

Leiserowitz, Anthony, Edward Maibach, Connie Roser-Renouf, and Peter Howe. 2012. *Extreme Weather and Climate Change in the American Mind*. New Haven, CT: Yale Project on Climate Change Communication.

Leland, S., and K. Thurmaier. 2009. "City-County Consolidation: Do Governments Actually Deliver on Their Promises?" Paper presented at the Urban Affairs Association Conference, Chicago, March 6–8.

Levin, Ben. 2011. "Comparing Canada and the U.S. on Education." *Education Week*, April 4.

Lewyn, Michael. 2010. "Sprawl in Canada and the United States." Draft LLM thesis, University of Toronto.

———. 2012. "Sprawl in Canada and the United States." *Urban Lawyer* 44, no. 1: 85–133.

Leyden, K. M. 2003. "Social Capital and the Built Environment: The Importance of Walkable Neighborhoods." *American Journal of Public Health* 93, no. 9: 1546–51.

Lipset, Seymour Martin. 1990. *Continental Divide: The Values and Institutions of the United States and Canada*. New York: Routledge.

Litman, Todd. 2014a. *Introduction to Multi-Modal Transportation Planning*. Victoria, BC: Victoria Transport Policy Institute.

———. 2014b. *Transit Price Elasticities and Cross-Elasticities*. Victoria, BC: Victoria Transport Policy Institute.

———. 2015. *Evaluating Transportation Land Use Impacts*. Victoria, BC: Victoria Transport Policy Institute.

Lopez, Russ. 2004. "Urban Sprawl and Risk for Being Overweight or Obese." *American Journal of Public Health* 94, no. 9: 1574–79.

Maibach, Edward, Connie Roser-Renouf, Emily Vraga, Brittany Bloodhart, Ashley Anderson, Neil Stenhouse, and Anthony Leiser. 2013. *A National Survey of Republicans and Republican-Leaning Independents on Energy and Climate Change*. New Haven, CT: Yale Project on Climate Change Communication.

Malcolm, Andrew H. 2005. *The Canadians.* New York: Times Books.

Margo, Robert. 1992. "Explaining the Postwar Suburbanization of the Population in the United States: The Role of Income." *Journal of Urban Economics* 31: 301–10.

McIntyre, Linda. 2013. "Agenda 21 and the Planning Commission." *Commissioner,* Summer, 1–2.

Mees, Paul. 2010. *Transport for Suburbia: Beyond the Automobile Age.* London: Earthscan.

Mercer, John, and Kim England. 2000. "Canadian Cities in Continental Context: Global and Continental Perspectives on Canadian Urban Development." In *Canadian Cities in Transition,* 2nd ed., edited by Trudi Bunting and Pierre Filion, 55–75. Don Mills, ON: Oxford University Press.

Mieszkowski, Peter, and Edwin Mills. 1993. "The Causes of Metropolitan Suburbanization." *Journal of Economic Perspectives* 7: 135–47.

Miller, Jason 2005. "Another Greyfield Gone: Belmar in Lakewood, Colorado." *Town Paper,* Fall.

Millward, Hugh. 2002. "Peri-urban Residential Development in the Halifax Region 1960–2000: Magnets, Constraints and Planning Policies." *Canadian Geographer* 46, no. 1: 33–47.

Miron, John. 2003. "Urban Sprawl in Canada and America: Just How Dissimilar?" Paper presented at the annual meeting of the Association of American Geographers, New Orleans, LA, March.

Moore, T., P. Thornes, and B. Appleyard. 2007. *The Transportation/Land Use Connection.* Planning Advisory Service Report No. 546/547. Chicago: American Planning Association.

Nelson, Arthur. 2010. "Resetting the Demand for Multifamily Housing: Demographic and Economic Drivers to 2020." Presentation to National Multi Housing Council.

———. 2013. *Reshaping Metropolitan America: Development Trends and Opportunities to 2030.* Washington, DC: Island Press.

Nelson, Arthur, Liza Bowles, Julian Juergensmeyer, and James Nicholas. 2008. *A Guide to Impact Fees and Housing Affordability.* Washington, DC: Island Press.

Neuharth, Al. 2006. "Traveling Interstates Is Our Sixth Freedom." *USA Today,* June 22.

Newman, Peter, and Jeffrey Kenworthy. 1989. *Cities and Automobile Dependence.* Brookfield. VT: Gower Technical Press.

———. 1999. *Sustainability and Cities: Overcoming Automobile Dependence.* Washington, DC: Island Press.

———. 2015. *The End of Automobile Dependence: How Cities Are Moving beyond Car-Based Planning*. Washington, DC: Island Press.

Nivola, Pietro S. 1999. *Laws of the Landscape: How Policies Shape Cities in Europe and America*. Washington, DC: Brookings Institution Press.

Norquist, John. 2014. "Roadblock on Main Street: How Federal Finance Policies Run Downtowns Out of Business." *American Conservative*, November/December.

Norris, Nathan. 2012. "Why Gen Y Is Causing the Great Migration of the 21st Century." *PlaceShakers and Newsmakers*, April 9.

Northwest Environment Watch and Smart Growth BC. 2002. *Sprawl and Smart Growth in Greater Vancouver: A Comparison of Vancouver, British Columbia, with Seattle, Washington*. Seattle, WA: Northwest Environment Watch.

Ontario Ministry of Infrastructure. 2013. *Growth Plan for the Greater Golden Horseshoe 2006, Office Consolidation 2013*. Toronto: Province of Ontario.

Oram, Richard. 1980. "The Role of Subsidy Policies in Modernizing the Structure of the Bus Transit Industry." *Transportation* 9, no. 4: 333–53.

Orfield, Myron. 1998. *Metropolitics: A Regional Agenda for Community and Stability*. Washington, DC: Brookings Institution Press.

Ostrom, V., C. Tiebout, and R. Warren. 1961. "The Organization of Government in Metropolitan Areas: A Theoretical Inquiry." *American Political Science Review* 55: 831–42.

Owen, David. 2010. *Green Metropolis: Why Living Smaller, Living Closer, and Driving Less Are the Keys to Sustainability. New York: Riverhead Books.*

Padova, Allison. 2006. *Federal Participation in Highway Construction and Policy in Canada*. Ottawa: Library of Parliament.

Parkinson, Giles. 2013. "HSBC: World Is Hurtling towards Peak Planet." *RenewEconomy* March 27.

Patriquin, Martin. 2014. "The Epic Collapse of Quebec Separatism." *Maclean's*, April 11.

Paulsen, Kurt. 2014. "Geography, Policy or Market? New Evidence on the Measurement and Causes of Sprawl (and Infill) in U.S. Metropolitan Regions." *Urban Studies* 51, no. 12: 2629–45.

Peck, Jamie. 2005. "Struggling with the Creative Class." *International Journal of Urban and Regional Research* 29, no. 4: 740–70.

Peden, Margie, Richard Scurfield, David Sleet, Dinesh Mohan, Adnan A. Hyder, Eva Jarawan, and Colin Mathers, eds. 2004. *World Report on Road Traffic Injury Prevention: Special Report for World Health Day on Road Safety*. Geneva: World Health Organization.

Pendall, Rolf, and Christopher R. Hayes. 2014. *Driving to Opportunity: Understanding the Links among Transportation Access, Residential Outcomes, and Economic Opportunity for Housing Voucher Recipients.* Washington, DC: Urban Institute.

Pendall, Rolf, Robert Puentes, and Jonathan Martin. 2006. *From Traditional to Reformed: A Review of the Land Use Regulations in the Nation's 50 Largest Metropolitan Areas.* Washington, DC: Brookings Institution Press.

Perl, Anthony, and John Pucher. 1995. "Transit in Trouble? The Policy Challenge Posed by Canada's Changing Urban Mobility." *Canadian Urban Transit Association Forum*, June 5, 22–24.

Persky, Joseph, and Wim Wiewel. 2000. *When Corporations Leave Town.* Detroit: Wayne State University Press.

Peterson, Paul. 1981. *City Limits.* Chicago: University of Chicago Press.

Pew Research Center. 2012. "The Lost Decade of the Middle Class." *Social and Demographic Trends*, August 22.

Pucher, J. 1994. "Public Transport Developments: Canada vs. the United States." *Transportation Quarterly* 48, no. 1: 65–78.

Pucher, John, and Ralph Buehler. 2006. "Why Canadians Cycle More Than Americans: A Comparative Analysis of Bicycling Trends and Policies." *Transport Policy* 13, no. 3: 265–79.

———. 2007. "Cycling in Canada and the United States: Why Canadians Are So Far Ahead." *Plan*, Spring/Summer, 13–17.

———. 2012. "International Overview." In *City Cycling*, edited by John Pucher and Ralph Buehler, 9–30. Cambridge, MA: MIT Press.

———, eds. 2012. *City Cycling.* Cambridge, MA: MIT Press.

Pucher, John, and Christian Lefevre. 1996. *The Urban Transport Crisis in Western Europe and North America.* London: Macmillan.

Raad, Tamim. 1998. "The Car in Canada: A Study of Factors Influencing Automobile Dependence in Canada's Seven Largest Cities 1961–1991." Master's thesis, University of British Colombia.

Rae, Douglas W. 2003. *City: Urbanism and Its End.* New Haven, CT: Yale University Press.

Reardon, Sean F., and Kendra Bischoff. 2011. "Growth in the Residential Segregation of Families by Income." Providence, RI: US 2010 Project, Russell Sage Foundation and Brown University.

Reese, Laura, and Amy Malmer. 1994. "The Effects of State Enabling Legislation on Local Economic Development Policies." *Urban Affairs Quarterly* 30, no. 1: 114–35.

Richmond, J. E. D. 1998. *New Rail Transit Investments: A Review.* Cambridge, MA: John F. Kennedy School of Government, Harvard University.

Rosenfeld, R. A., and L. A. Reese. 2004. "Local Government Amalgamation from the Top Down." In *City-County Consolidation and Its Alternatives: Reshaping the Local Government Landscape*, edited by J. B. Carr and R. C. Feiock, 219–46. Armonk, NY: M. E. Sharpe.

Ross, Bernard, and Myron A. Levine. 2012. *Urban Politics: Cities and Suburbs in a Global Age*. Armonk, NY: M. E. Sharpe.

Rothblatt, Donald. 1994. "North American Metropolitan Planning: Canadian and US Perspectives." *American Planning Association Journal* 60, no. 4: 501–520.

Rusk, David. 1993. *Cities without Suburbs*. Washington, DC: Woodrow Wilson Center Press.

Sancton, Andrew. 2005. "The Governance of Metropolitan Areas in Canada." *Public Administration and Development* 25, no. 4: 317–27.

Schimek, Paul. 1996. "Automobile and Public Transit Use in the United States and Canada: Comparison of Postwar Trends." *Transportation Research Record* 1521: 3–11.

———. 1997. "Understanding the Relatively Greater Use of Public Transit in Canada Compared to the USA." PhD diss., Urban Studies and Planning Department, Massachusetts Institute of Technology.

Secretary of Defense. 2014. *Quadrennial Defense Review*. Washington, DC: US Department of Defense.

Sharpe, Andrew, and Evan Capeluck. 2012. *Canadians Are Happy and Getting Happier: An Overview of Life Satisfaction in Canada, 2003–2011*. Ottawa: Center for the Study of Living Standards.

Shin, Yong Eun, V. R. Vuchic, and E. Christian Bruun. 2009. "Land Consumption Impacts of a Transportation System on a City: An Analysis." *Transportation Research Record* 2110: 69–77.

Shoup, Donald. 1999. "Instead of Free Parking." *Access* 15, no. 2: 6–9.

Skelly, Michael J. 1995. *The Role of Canadian Municipalities in Economic Development*. Toronto: ICURR Press.

Skogan, Wesley. 1990. *Disorder and Decline: Crime and the Spiral of Decay in American Neighborhoods*. Berkeley: University of California Press.

Sohmer, Rebecca, and Robert Lang. 2001. *Downtown Rebound*. Washington, DC: Fannie Mae Foundation / Brookings Institution, Center on Urban and Metropolitan Policy.

Sorensen, A., R. P. Greene, and K. Russ. 1997. *Farming on the Edge*. DeKalb, IL: American Farmland Trust.

Speck, Jeff. 2012. *Walkable City: How Downtown Can Save America, One Step at a Time*. New York: Farrar, Straus and Giroux.

Standing Committee on Human Resources, Skills and Social Development and the Status of Persons with Disabilities, House of Commons. 2010. *Federal Poverty Reduction Plan: Working in Partnership towards Reducing Poverty in Canada.* Ottawa: Parliament of Canada.

Stephenson, Gordon. 1957. *A Redevelopment Study of Halifax, Nova Scotia.* Halifax, NS: City of Halifax.

Stevens, John M., and Robert P. McGowan. 1987. "Patterns and Predictors of Economic Development Power in Local Government: A Policy Perspective on Issues in One State." *Policy Studies Review* 6 (February): 554–68.

Stoll, Michael. 2007. "Spatial Mismatch and Job Sprawl." In *The Black Metropolis in the Twenty-First Century: Race, Power, and Politics of Place,* edited by R. D. Bullard, 127–48. Lanham, MD: Rowman and Littlefield.

Sturm, R., and D. Cohen. 2004. "Suburban Sprawl and Physical and Mental Health." *Public Health* 118, no. 7: 488–96.

Taber, Jane. 2014. "Interchange Demolition Plan gives Halifax Renewal Rare Second Chance." *Globe and Mail,* May 8.

Tal, Gil, and Susan Handy. 2011. "Measuring Non-motorized Accessibility and Connectivity in a Robust Pedestrian Network." Final Research Report So2-2. Sustainable Transportation Center, University of California, Davis.

Taylor, Zachary. 2008. *Shaping the Toronto Region, Past, Present, and Future.* Toronto: Neptis.

———. 2015. "The Politics of Metropolitan Development: Institutions, Interests, and Ideas in the Making of Urban Governance in the United States and Canada, 1800–2000." PhD diss., Graduate Department of Political Science, University of Toronto.

Taylor, Zachary, and Neil Bradford. 2015. "The New Localism: Canadian Urban Governance in the Twenty-First Century." In *Canadian Cities in Transition: Perspectives for an Urban* Age, edited by P. Filion, M. Moos, T. Vinodrai, and R. Walker, 194–208. Don Mills, ON: Oxford University Press.

Taylor, Zachary, Marcy Burchfield, and Anna Kramer. 2014. "Alberta Cities at the Crossroads: Urban Development Challenges and Opportunities in Historical and Comparative Perspective." School of Public Policy Research Papers, University of Calgary.

Taylor-Vaisey, Nick. 2012. "Where Canadian Criminals Go to Play: A Look at the Cities with the Most Lawbreakers." *Maclean's,* November 29.

Teaford, Jon. 1986. *The Twentieth-Century American City.* Baltimore: Johns Hopkins University Press.

Teixeira, Carlos, Wei Li, and Audrey Kobayashi, eds. 2011. *Immigrant Geographies of North American Cities.* Don Mills, ON: Oxford University Press.

Tennant, Paul, and David Zirnhelt. 1973. "Metropolitan Government in Vancouver: The Politics of Gentle Imposition." *Canadian Public Administration*, 16 (Spring): 124–38.

Thomas, David M., and David N. Biette, eds. 2014. *Canada and the United States: Differences That Count*. 4th ed. Toronto: University of Toronto Press.

Tomalty, Ray. 2007. *Innovative Infrastructure Financing Mechanisms for Smart Growth.* Vancouver: Smart Growth BC.

Tomalty, Ray, and Don Alexander. 2005. *Smart Growth in Canada: Implementation of a Planning Concept.* Ottawa: Canadian Mortgage and Housing Corporation.

Tomalty, Ray, and Murtaza Haider. 2008. *Housing Affordability and Smart Growth in Calgary.* Calgary, AB: City of Calgary.

Torrens, Paul, and Marina Alberti. 2000. *Measuring Sprawl.* London: Centre for Advanced Spatial Analysis.

Town of Huntington, New York. 2008. *Horizons 2020: A Comprehensive Plan Update.*

TransLink. N.D. "Making the Connection to a World of Choice."

Transportation Research Board. 1997. *Special Report 251: Toward a Sustainable Future: Addressing the Long-Term Effects of Motor Vehicle Transportation on Climate and Ecology.* Washington, DC: National Research Council.

Transportation Research Board, National Research Council. 2001. *Making Transit Work: Insight from Western Europe, Canada, and the United States—Special Report 257. Washington, DC: National Academies Press.*

Tu, Charles C., and Mark J. Eppli. 2001. "An Empirical Examination of Traditional Neighborhood Developments." *Real Estate Economics* 29, no. 3: 485–501.

Turcotte, Martin. 2008. "The City/Suburb Contrast: How Can We Measure It?" *Canadian Social Trends* 85(Summer).

Urban Land Institute. 2012. *Shifting Suburbs: Reinventing Infrastructure for Compact Development*. Washington, DC: Urban Land Institute.

Voith, Richard. 1999. "Does the Federal Tax Treatment of Housing Affect the Pattern of Metropolitan Development?" *Federal Reserve Bank of Philadelphia Business Review*, 3–16.

Vojnovic, Igor. 1998. "Municipal Consolidation in the 1990s: An Analysis of British Columbia, New Brunswick, and Nova Scotia." *Canadian Public Administration* 41, no. 2: 239–83.

———. 2000. "The Transitional Impacts of Municipal Consolidation." *Journal of Urban Affairs* 22, no. 4: 385–417.

Vornovytskyy, Marina, Alfred Gottschalck, and Adam Smith. N.D. *Household Debt in the U.S., 2000 to 2011.* Washington, DC: US Census Bureau.

Wachs, M. 1989. "U.S. Transit Subsidy Policy: In Need of Reform." *Science*, June, 1545–49.

Walks, R. Alan, and Larry S. Bourne. 2006. "Ghettos in Canada's Cities? Racial Segregation, Ethnic Enclaves and Poverty Concentration in Canadian Urban Areas." *Canadian Geographer 50,* no. 3: 273–97.

Williamson, Thad. 2010. *Sprawl, Justice, and Citizenship: The Civic Costs of the American Way of Life.* New York: Oxford University Press.

Winters, M., G. Davidson, D. Kao, and K. Teschke. 2011. "Motivators and Deterrents of Bicycling: Comparing Influences on Decisions to Ride." *Transportation* 38: 153–68.

Wolff, Edward N. 2014. *Household Wealth Trends in the United States, 1962–2013: What Happened over the Great Recession?* NBER Working Paper No. 20733. Cambridge, MA: National Bureau of Economic Research.

Wood, Robert C. 1958. *Suburbia: Its People and Their Politics.* Boston: Houghton Mifflin.

Wright, D. 1988. *Understanding Intergovernmental Relations.* Pacific Grove, CA: Brooks/Cole.

Yeates, Maurice, and Barry Garner. 1971. *The North American City.* New York: Longman.

Index

Page numbers followed by "f" and "t" indicate figures and tables.

Island Press | Board of Directors